Astronomy & Telescopes

A Beginner's Handbook

Other TAB books by Robert J. Traister

Astronomy & Telescopes
A Beginner's Handbook

by Robert J. Traister & Susan E. Harris

TAB TAB BOOKS Inc.

BLUE RIDGE SUMMIT, PA. 17214

FIRST EDITION
SECOND PRINTING

Copyright © 1983 by TAB BOOKS Inc.
Printed in the United States of America

Color photos courtesy of Celestron International, Torrance, California, U.S.A.

Library of Congress Cataloging in Publication Data

Traister, Robert J.
Astronomy and telescopes.

Includes index.
1. Astronomy—Observers' manuals. 2. Telescope—
Handbooks, manuals, etc. I. Harris, Susan E.
(Susan Elizabeth), 1949- .II. Title.
QB64.T7 1983   522'2   82-19346
ISBN 0-8306-0419-7
ISBN 0-8306-1419-2 (pbk.)

Front cover: Lunar eclipse photographed with Celestron 8 telescope.
Back cover: Solar eclipse photographed with Celestron 5 telescope.
Photos courtesy of Celestron International, Torrance, California, U.S.A.

Contents

Introduction

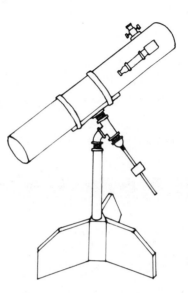

ASTRONOMY IS THE STUDY OF CELESTIAL bodies in the heavens. Galileo, Leonardo da Vinci, and Sir Isaac Newton pioneered astronomy and developed the basis for astrophysics.

Astronomy was for a select few in those days, but amateur astronomy is very popular today. High-quality instruments are available at moderate prices to make astronomy an affordable hobby for the individual or family.

This book will take you from the beginnings of astronomy to the present. The book is not some highly technical text designed only for those with years of experience in astronomy, astrophysics, and highly complex math.

If you want to spend thousands of dollars, some of the products described will be of interest. If your budget permits only a small expenditure of $100 or so, you will also find what you need.

There is nothing quite so relaxing as an evening spent viewing the heavens. As you gain more experience in astronomy, you will have a better perception of the universe.

To Wayne Warren Williams—a friend, supporter, and the one who is bound to reach astronomical heights in everything he does.

Chapter 1

Astronomy: from Birth to Present

THE EGYPTIANS, INCAS, AND OTHER ANCIENT civilizations were very interested in astronomy. Modern scholars still marvel at their precise calculations. Most early observations were accomplished without optical instruments and were based on a mathematical system for calculating planetary orbits and the characteristics of other celestial bodies.

Even primitive man used movements of celestial bodies to measure time, particularly regarding crop planting. Most early measurements were probably directly related to the moon's motion, which resulted in the establishment of planting cycles and a time system. There was really no scientific study involved. Although these early observations did provide valuable information, most ancient civilizations did not contemplate the whys and hows of later years.

The Egyptians established the first recorded calendar based on the sun's cycle. The Babylonians provided the *sexagesimal* system, which established that an hour consisted of 60 minutes.

A monument to early man that remained a mystery for many years is *Stonehenge,* England. It was originally thought that these immense stones, neatly arranged in a circle, were a temple or religious ceremonial site. Recent studies show that their layout is positioned to indicate the summer solstice or the day when the sun reaches its northernmost position in the sky.

These earlier people thought the earth was a flat sphere and that celestial bodies sunk into the oceans and then surfaced again after receiving this "bath." This theory persisted until observers in Greece around the sixth century B.C. noted that the earth was spherical due to the shadow it cast on the moon during an eclipse. Further calculations during the next few hundred years supported this theory and raised more questions as to what strange phenomenon held the earth in place.

ASTRONOMERS

In the second century, A.D. Ptolemy theorized that the earth was the center of the universe and the moon, sun, and other bodies were in a structured pattern around it. He presented orbital paths of the planets and stars as perfect circles, each a little larger than the one inside it (Fig. 1-1). This theory remained intact for 13 centuries before it was seriously questioned. Churches favored Ptolemy's theory because it placed man in an exalted, godlike position.

Copernicus, Galileo, and Tycho Brahe

Copernicus was the first astronomer to seriously refute Ptolemy's theory. His theory was that

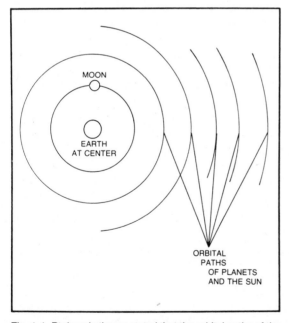

Fig. 1-1. Ptolemy's theory stated that the orbital paths of the known stars and planets formed perfect circles, with each one a little larger than the successive body.

the earth rotated on an axis around the sun. He also determined the orbital periods of the five planets charted at that time. His ideas were presented in a book that was unfortunately withheld by the Catholic church for many years after his death.

Galileo and Johann Kepler attempted to obtain scientific proof that would substantiate Copernicus's claims. Although Galileo did not actually invent the telescope, he made it a popular astronomical instrument. Galileo's and Kepler's observations through a primitive telescope proved that the sun was the center of our solar system.

Tycho Brahe used primitive instruments to chart the positions of the sun, moon, planets, and stars. His observations laid the groundwork for Kepler.

Kepler

Because Kepler had an eye disease, he was unable to do any accurate observational work. He studied the works of Tycho to discover any discrepancies that would disprove his theories. This research led to Kepler's planetary motion laws that depicted the earth and other planets' orbits as elliptical rather than perfect circles. He determined that the speed with which a planet moved in its orbit was directly related to its distance from the sun. He also thought the sun exerted some force that controlled the motion of bodies revolving around it.

Kepler had some highly unusual ideas. One theory stated that each planet emitted a harmonic sound, while another stated that there was possibly a demon inside each planet. These bizarre theories are probably attributed to his belief in the occult and supernatural forces. He was a highly superstitious man who was well known for his astrological chartings.

Galileo's observation that the earth was not the center of the universe became more widely accepted during this period. His experiments with the laws of motion were later expanded by Sir Isaac

Newton and explained the planets' and other bodies' positions in relation to the sun.

Newton

Newton believed that a body would remain at rest or continue to move in the same direction at the same speed unless acted upon by another force. In other words, every action has an equal and opposite reaction. He could accurately determine planets' orbits, and those of other celestial bodies, based on the knowledge that gravitational pull held each of them in their positions. As each planet exerts its own centrifugal force in an opposite direction, the sun exerts an equal amount of gravitational force to maintain an orbital balance.

Newton made some changes in telescope design that are still with us today. The end product bears his name—the Newtonian or *reflector* telescope. In early days all telescopes were classified as *refractors*, using two or more lenses in longitudinal alignment. This design has an inherent problem of chromatic aberration caused by the differences in refraction of the spectrum's colored rays.

The Newtonian reflector used a reflecting mirror instead of a lens as the main collector of light. This curved mirror focused the gathered light onto a single lens. Chromatic aberation does not take place, as all wavelengths of light reflected by the mirror are equally bent due to its *parabolic* curvature.

Newton's contributions were not widely accepted until after his death in 1727. His theories regarding planetary orbits and his experiments with light led to the invention of a telescope that is still widely used today.

Herschel, Kirchhoff, and Bunsen

Walter Herschel discovered the planet Uranus about 150 years after Newton's contributions. He advanced Newton's earlier theories regarding light.

Newton's studies indicated that white light, when passed through a prism, breaks up into a spectrum of colors. The colors were determined by the speed with which the waves passed through the prism. Herschel discovered that there is invisible infrared radiation beyond the spectrum. Further studies have led to the discovery of the electromagnetic specturm, radio waves, and gamma and X rays. In the late 1800s two German physicists, Kirchhoff and Bunsen, learned that the colors of an element will vary. They were able to accurately determine the composition of planets, the sun, moon, and other celestial bodies. *Spectroscopy* has enabled scientists and astronomers to learn about the atmospheric variations in our solar system and identify the gases in the atmosphere. Special telescopes provide information through the use of infrared and radio waves.

SATELLITES

The first man-made *satellite*, Sputnik I, was successfully launched in 1957 by the Russians. Satellites are not only difficult to locate once in orbit, but they are affected by fluctuations in gravity, radiation from the sun, and frictional effects of the earth's atmosphere. A system had to be developed whereby these satellites could be tracked so that the information they provide could be used. The first tracking systems were extremely large cameras that worked in conjunction with a radio tracking system. This eventually led to the use of laser beams for satellite tracking at greater accuracy.

Satellites are commonly used for transmitting sound waves. This expanded communication system has brought us closer to our neighbors. Returned satellites provided scientists with the information they needed to know before attempting a manned mission. By measuring the temperatures over the entire surface of the satellite, scientists have determined the amount of stress that persons inside a space capsule would have to withstand, and what adjustments would be necessary to enable the eventual launching of man into space. The inter- communication system necessary for a manned flight was developed through the use of satellites by means of special equipment.

Chapter 2

Telescopes: Refractors, Reflectors, and Complex Designs

THE INVENTION OF THE TELESCOPE IN THE seventeenth century heralded the beginning of a new science. Although these first instruments were very primitive compared to those in use today, the same principles have remained.

The advances in astronomy paralleled to a great extent the advances made in the telescope. As the instrument itself was improved, it was possible to see more clearly and much further into the outer reaches of the universe. Astronomers could study the surfaces and atmospheres of other celestial bodies.

Before discussing various telescopes, it is necessary to dispense with some myths surrounding these instruments. The amateur astronomer should have a basic understanding of some common terms and principles involved in the manufacture of telescopes and their components.

Magnification or power is perhaps the most widely misunderstood telescope characteristic. In an astronomical telescope the power of the instrument is variable depending upon the eyepiece used. Power is computed by dividing the focal length of the primary mirror (or objective lens in the case of a refracting telescope) by the eyepiece's focal length. A 6-inch f/8 telescope with a focal length of 48 inches operates at a power of 135 × when used with a 9-mm eyepiece. An 8-inch f/6 telescope has a focal length of 48 inches and also yields 135 × when used with a 9-mm eyepiece. The 8-inch telescope at 135 × shows far more detail than the 6-inch telescope at 135 ×, because the larger aperture results in 78 percent more light entering the telescope. The key to the observation of fine details is not power but light-gathering capability. The larger the telescope, the more light enters the system. Telescopes with higher powers can be practical and useful.

To determine the focal ratio or f/ratio of a

telescope, you must know the diameter of the object glass or mirror relative to its focal length. A 6-inch diameter telescope with a 48-inch focal length implies a focal ratio of f/8. The focal length of a telescope is the distance between the lens (or mirror) and the point where the light rays are focused (the focal point).

APERTURE

There are three basic telescope designs in use today: reflectors, refractors, and combination instruments. There is much debate among astronomers and educators as to which performs best. All designs will perform satisfactorily if properly and responsibly made. Each design has its own merits and drawbacks. Celestron, one of the leading manufacturers of telescopes, states that a clear aperture is far more significant to performance than any design feature. Optical theory firmly establishes the limits of light grasp and resolution for a given aperture.

☐ Resolution is a direct function of the aperture.
☐ Light grasp is proportional to the square of the aperture.
☐ Airy disk brilliance (the brightness of a point source stellar image) is proportional to the fourth power of the aperture.

When you double a telescope's aperture, you increase its resolving power by a factor of two and boost its light-gathering power by a factor of four. You also reduce the area of the Airy disk by a factor of four. This results in a 16-fold gain in image brilliance.

A globular cluster such as M-13 is totally amorphous with a 4-inch telescope at 150 ×, no matter how good the viewing conditions are. The stars are separated twice as well with an 8-inch telescope at the same power. The stars are 16 times more brilliant. The cluster is resolved to the core, even during poor viewing periods.

The same general principle applies when viewing the detailed structure of planets, nebulae, and galaxies. When you increase the aperture of a telescope, you dramatically increase the amount of visible detail and contrast. It makes little difference whether the comparison telescopes are refractors, reflectors, or catadioptrics.

One point of debate is the contention that smaller instruments can split the closest doubles and reveal the smallest planetary detail with greater ease than larger, professional telescopes. Atmospheric scintillation will not permit the larger telescopes to achieve the resolution that their larger apertures make possible. When turbulence is great, a 4-inch telescope will show the same amount of planetary detail that is visible through the Mount Wilson 100-inch reflector or the Lick 36-inch refractor under the same conditions. On nights of gross turbulence at high magnification, the Celestron 5 will sometimes render a superficially steadier view of some bright objects than the Celestron 8 or 14. This is never the case with low-contrast subjects such as nebulae and star clusters. The Celestron people believe that the smaller instrument is incapable of resolving turbulence to the extent of the larger. The same effect may be achieved by lowering the magnification of the larger instrument so that turbulence remains unresolved. It is possible to enjoy the benefits of substantial aperture, much greater image contrast, and brilliance. The larger the aperture of the telescope at any given magnification, the better the image will be.

REFRACTORS

The *refracting* telescope is a mechanically simple, rugged, and maintenance-free instrument (Fig. 2-1). The closed tube design protects its optics and eliminates image-degrading air currents inside the tube. Simple refracting telescopes are relatively small (2.4-inch to 3.1-inch) and readily available from department stores. Some ads claim these instruments allow 600 ×, which is just empty they suffer from chromatic aberration, which is seen as fringes of color at the image's outer edges. This defect has been corrected to some extent in modern refracting telescopes. Chromatic aberration is harmful in studying the colors of the stars and planets.

An advantage of the refractor is that aperture for aperture (same focal lengths), it offers somewhat less light loss. Its lack of a secondary mirror obstruction (or diagonal obstruction) in very small sizes generally gives it a slight edge in performance over Newtonians and catadioptric telescopes. Refractors usually come in long focal lengths (f/15 or slower) that provide excellent views of lunar and planetary subjects and also double stars.

Refractors can be used for terrestrial observing, but the near focus usually prohibits detailed studies of subjects closer than 100 feet. The focal lengths are usually long (f/15 or slower), which makes photography more difficult. Refractors are quite bulky and not very portable, particularly when the instrument reaches an aperture of 3 or more inches. A quality 5-inch refractor will probably weigh several hundred pounds.

Refracting telescopes are easy to use when compared to a comparable reflector type. You may need a special viewing stand for a large reflecting telescope to reach the eyepiece, which is mounted to the front portion of the instrument. A refractor uses a rear-mounted eyepiece that is the closest point to the ground when the instrument is oriented toward the sky. The refractor is a sealed instrument. The inner tube assembly is not as subject to dust contamination as is the reflector, which is open to the elements at one end. The outer surfaces of the lenses are the only areas that need maintenance. Dust should be removed with a proper cleaning cloth.

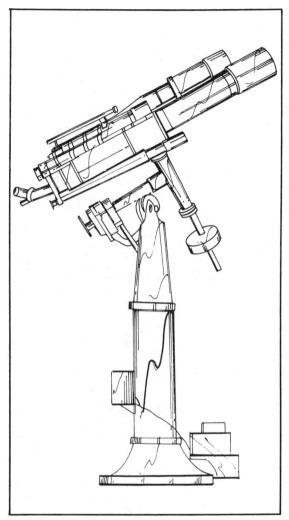

Fig. 2-1. A refracting telescope is a relatively maintenance-free instrument due in part to its closed tube design.

magnification. The maximum useful power is 60 times the distance of the lens. A 3-inch telescope has a maximum power of 180 × (3×60).

Disadvantages and Advantages

A quality 5-inch refractor may cost almost $10,000. Another disadvantage of refractors is that

Optical Operation of the Refractor

Figure 2-2 is an internal diagram of a simple refracting telescope. Note that the objective lens is

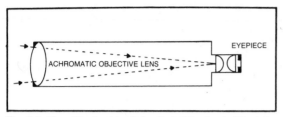

Fig. 2-2. The objective lens is located at the front of the tube assembly.

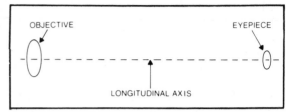

Fig. 2-4. Two Lenses are aligned along the same axis in a refractor.

placed at the front portion of the tube assembly. When light passes through this lens, it is bent inward until it is focused at the tube's far end. The eyepiece receives the image at this focal point, allowing it to be focused on the human eye.

A refracting telescope can be made from two lenses. A tube housing is not necessary. Figure 2-3 shows this principle. You can perform this experiment with two common magnifying glasses. Hold one out in front of you as far as possible, then place the other one in line with the first. By moving the second glass toward your viewing eye, you can achieve the magnification possible with the type of lenses being used. This system is almost exactly like the first telescope ever made.

The refracting telescope consists of two lenses centered on the same axis (Fig. 2-4). The main lens or objective is mounted at the instrument's far end. The diameter of this lens in telescopes designed for amateur astronomers may be from 2 or 3 inches to 5 or 6 inches. Scientific refractors may contain lenses of 1 foot or more in diameter. The largest refractor in the world contains an aperture of more than 3 feet. This instrument is more than 60 feet long and weighs many tons. The focal length of a typical refractor is usually held to about 15 times the lens diameter.

The objective lens focuses the image of an object lying far away to a specific point along its axis. The objective lens captures the light from the object to be viewed and then focuses it to a specific and microscopic point. All of the gathered light is concentrated at a definitive point.

The eyepiece lens acts like the objective lens in some ways. It focuses the tiny point of light on the human eye. The magnifying power of this telescope is equal to the focal length of the objective lens divided by the focal length of the eyepiece. For example, with an objective of a 4-foot focal length coupled with an eyepiece with a 1-inch focal length, the magnifying power would be equal to 1 inch divided into 48 inches (4 feet equals 48 inches), or a magnifying power of 48 ×. The image would be magnified 48 times. In refracting telescopes the focal length is usually the physical length of the tube and eyepiece assembly. If your refractor measures

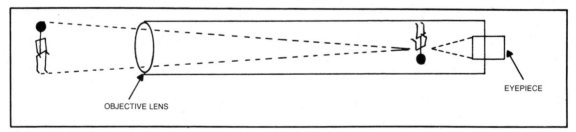

Fig. 2-3. The basic principle of a refracting telescope is quite simple.

60 inches in length from objective to eyepiece, the focal length will be approximately 60 inches or a little more than 1500 mm. You should compute the focal length in millimeters, because the focal lengths of various lenses are almost always given in millimeters. To determine the operational power of your refractor telescope with various lenses, divide the objective focal length or the physical length of the instrument when the former is not given by the lens focal length. A telescope with a 1524-mm focal length (about 60 inches) will produce a magnification of approximately 54 × when a 28-mm eyepiece is used (1544 divided by 28). When used with a 12-mm eyepiece, this telescope will produce a magnification of approximately 120 ×.

When you decrease eyepiece focal length, you increase magnification power when both eyepieces are applied to the same instrument. When you increase power, though, you also decrease the effective light-gathering ability of the instrument. The image viewed with a 28-mm lens will be brighter and sharper than that produced with a 12-mm eyepiece.

There are upper and lower limits of magnification for any telescope. The upper limit is reached when the image is so dim as to be completely useless for the recovery of information. The upper limit of magnification can be achieved only when the air is very steady, and it is about 60 times per inch of aperture. A telescope with a 3-inch aperture would have a maximum usable magnification of 180 ×. There is a sacrifice of image quality. The lower limit of magnification is approximately four times per inch and, for the telescope with a 3-inch aperture, approximately 12 ×.

An eyepiece with a focal length of 8.5 mm would be the maximum that could be used with a 60-inch refractor for practical magnification, assuming that this refractor had a 3-inch aperture. The maximum referred to here involves magnification. An 8.5-mm focal length is the minimum focal length that can be used with this 3-inch telescope.

A lens with a focal length of 128 mm would be the minimum magnification that could be used with this instrument. This would produce a magnification power of 12 ×, which is four times the aperture length. Eyepiece lenses with focal lengths of 60 mm will be the largest encountered in practical applications.

As you increase the focal length of the eyepiece, the magnification properties are decreased. The brightness of the image increases. Clarity is usually better, and there is less atmospheric distortion when viewing celestial objects. Maximum clarity is obtained at minimum magnifying powers for individual instruments.

NEWTONIAN REFLECTOR

The most universally used telescope in the world today is the *Newtonian reflector* (Fig. 2-5). The Newtonian is still very cost effective, and professional quality instruments can be purchased for a fraction of the cost of more complex designs.

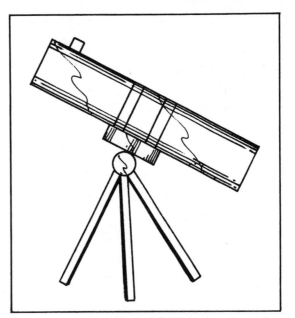

Fig. 2-5. A Newtonian reflector is both simple and versatile.

Figure 2-6 shows how a reflecting telescope works. Instead of passing the incoming light rays through a glass lens, the reflecting telescope bends the rays by reflecting them from the surface of a concave mirror. The only lens that is used in the Newtonian design is found in the eyepiece. Light from the object being viewed travels down the tube assembly where it strikes the primary mirror. The light is then reflected toward a focal point that lies just outside the tube assembly. The focused rays do not quite make it to this point. They are intercepted by a secondary diagonal mirror that again bends the rays at a 90-degree angle to the incoming light. The latter light rays are then brought into contact with the eyepiece, which magnifies the image and directs the light to the eye. Due to the bending of the light rays, the viewed image is seen upside down. The physical space occupied by the secondary diagonal mirror prevents a small portion of light from reaching the primary mirror and the eyepiece, but this loss is negligible.

The Newtonian telescope is not an inferior instrument. The largest telescope in the world, the 200-inch Hale telescope at Mount Palomar, is a Newtonian reflector. The aperture is 200 inches, and the primary mirror is 200 inches wide. The mirror is rated in tons and not pounds.

The Newtonian telescope is not sealed like the refractor. The open end at the tube's front will allow dust to accumulate on the mirrors. The tube is usually sealed with a protective covering when the instrument is not in use. Do not place any clear glass over the surface of this opening, because it will interface with the travel of the light rays. Air turbulence within the open-ended tube can put it out of commission. This open tube design exposes the delicate mirror surfaces to contaminants, which means they must be recoated periodically. In a Newtonian the diagonal obstructs some light, and its small size somewhat limits the working field.

You can purchase a 12 ½-inch Newtonian for approximately $2,000. Small 6-inch instruments can be purchased for $250 to $400. Most Newtonian optical systems are well-made. The main optical aberration is *coma*, which is a cometlike distortion of images that increases the closer an image is to the field's edge. Astigmatism and curvature of field are other problems. A Newtonian telescope may be used for terrestrial applications, but it is awkward. The near focus is seldom under a few hundred feet.

An advantage of the Newtonian over a refractor of the same aperture and f/number is that it has a

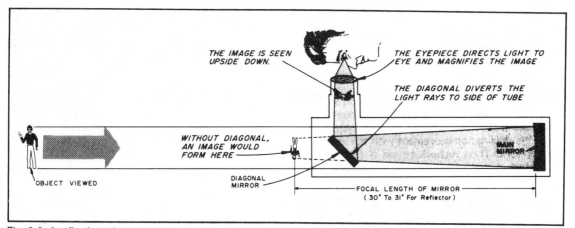

Fig. 2-6. A reflecting telescope utilizes mirrors rather than lenses (courtesy Edmund Scientific Company).

slightly shorter tube. This is not really significant, though, because a 6-inch Newtonian can weigh up to 100 pounds. The long tube of a Newtonian requires a large and massive mount to achieve a reasonable degree of stability. Depending upon the design and convenience of disassembly and reassembly, a Newtonian telescope may be impractical for remote viewing sessions.

Newtonians are available in short focal lengths (f/4 to f/8). These give excellent bright images for deep sky objects.

Magnification Power

Newtonian reflectors adhere to the same magnification principles as refracting telescopes. The focal length of the reflector is measured between the primary mirror, the secondary mirror, and the eyepiece. The average focal length will be just slightly longer than the length of the tube assembly. The aperture is the same as the primary mirror diameter. Figure 2-7 shows an 8-inch f/5 Newtonian reflector sold by Edmund Scientific. This instrument has a primary mirror that is 8 inches in diameter. The aperture is 8 inches or 203 mm. The focal length is about 40 inches or 1016 mm, which is just slightly longer than the length of the tube assembly. The maximum useful magnification will be 60 times the aperture dimensions or 480 ×. The minimum magnification will be four times the aperture diameter or 32 ×.

To achieve specific magnification powers, divide the focal length of the eyepiece into the focal length of the primary mirror, which is 1016 mm. A 28-mm eyepiece has a magnification power of 36 ×, 15-mm eyepiece 68 ×, and an 8-mm eyepiece has a magnification power of 127 ×. Eyepieces available for the average astronomer usually have minimum focal lengths of 4 mm. Using such an eyepiece with this instrument will result in a magnification of 254 ×. The maximum usable power is 480 × due to the aperture size.

You may wonder why a telescope was designed that could not conventionally take advantage of its maximum magnification abilities. You will find that this is often the case. This particular telescope has a relatively short focal length. It is designed for deep sky viewing, which means it must be able to capture a large portion of the sky. The low f/number indicates that its width of coverage or angle of arc is substantially wide. A telescope with a longer focal length using the same size aperture would have a higher f/number and take in a smaller portion of the sky.

Assume that this telescope is modified to provide for a longer focal length. This would mean a change in the design of the primary mirror and a lengthening of the tube assembly. If the telescope were to have a focal ratio of 10 or f/10, the focal length would have to be doubled, assuming that the

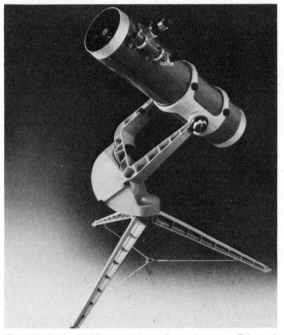

Fig. 2-7. An 8-inch f/5 Newtonian reflector (courtesy Edmund Scientific Company).

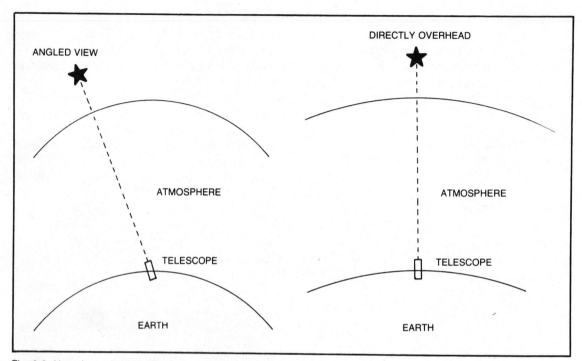

Fig. 2-8. Note the variation in the distance that light rays must travel under different viewing locations.

aperture remains 8 inches. Each of the lenses discussed earlier will now give double the magnification power. The maximum useful power remains the same regardless of focal length, as it is determined by the size of the aperture. A 4-mm eyepiece produced 254 × in the f/5 design, and the same lens would give a magnification of 508 ×—higher than the maximum useful magnification.

Focal Ratios

You may think that only telescopes with high focal ratios are worth buying. Nothing could be further from the truth. Maximum instrument power can only be used on rare occasions. This requires absolutely perfect atmospheric conditions. The image produced at the eyepiece will not be nearly as clear as that obtained at a lower power. Professional astronomers usually use the lowest possible

magnification in their observations, because they know that low power produces the clearest images. They resort to high power operation infrequently.

If you confine your observations only to the moon and the planets in our solar system, a high f/number may be desirable. If you want to view distant galaxies and star clusters, a telescope with a higher f/number may limit you to viewing only a portion of these celestial phenomena. This telescope will be more limited in viewing arc than one with a lower f/number. Most astronomers prefer instruments with low focal ratios, especially when dealing with apertures of the sizes most common to amateur astronomers. The Newtonian design can offer extremely low focal ratios. Most refracting telescopes have f/numbers of f/10, f/15, or higher. These higher ratios can be accomplished with a reflector design. Most large aperture amateur reflecting telescopes have f/numbers of between f/5 and f-8. Small aperture models in the 3 and 4-inch category will generally have f/numbers from f/8 to f/10.

There are mechanical disadvantages associated with viewing through a Newtonian reflector. The eyepiece is mounted to the "sky end" of the instrument. This is not a major problem even when using instruments with long focal lengths and tube sizes if viewing is restricted too near the horizon. This is not the ideal viewing angle, because the light must pass through much more atmosphere to reach the aperture than when viewing an object directly overhead.

The left-hand portion of Fig. 2-8 illustrates a telescope aimed at a point near the horizon. The light rays from the celestial object must pass through a much greater portion of the atmosphere to reach the aperture. The right-hand portion of Fig. 2-8 shows an overhead view. The shortest possible path exists between the incoming light and the telescope as referenced to the atmosphere. The earth's atmosphere contains dust, moisture, and other contaminants that can make astronomical observation almost impossible. When these contaminants are set into motion by strong atmospheric currents, the constant movement is almost impossible to contend with.

Many astronomy neophytes feel that the winter months would be best for celestial viewing, because winter skies often contain few clouds. The stars may seem much more sharply defined than in spring and summer months. Winter is often a very poor viewing time due to the increased atmospheric disturbances. You may recall noticing the pronounced twinkling of stars in the winter sky. The twinkling effect is created when incoming light rays are bent by atmospheric turbulence.

CASSEGRAIN REFLECTOR

The classical *Cassegrain reflector* telescope

shown in Fig. 2-9 is an improvement over the Newtonian telescope. It uses a short focal length parabolic mirror in conjunction with a small hyperbolic secondary mirror. This combination provides the same performance as the long tube Newtonian, but it does so in a far more compact size. A less massive mount is needed for the same degree of stability. The eyepiece is placed at a more convenient position. The disadvantages of the Cassegrain reflector telescope are the difficulty in parabolizing the short focal length primary mirror to the required degree of accuracy (much more costly than a Newtonian) and the open tube construction that exposes the mirrors to contamination. The Cassegrain reflector also suffers from coma to the same degree as the Newtonian.

The Cassegrain telescope is a reflecting type employing two curved mirrors to gain a long focal length. Light is beamed from the main mirror back up the tube to a second mirror, which beams it right back down again and through a hole in the center of the main mirror to the eyepiece.

Other than the true Cassegrain system, there are two recognized systems made—the *Dall-Kirkham* and the *Ritchey-Chretien*. The Dall-

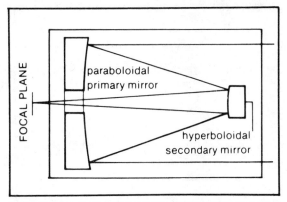

Fig. 2-9. The classical Cassegrain reflector telescope combines the features of both reflector and refractor designs (courtesy Celestron International, Torrance, California, U.S.A.).

Kirkham system is cheaper to make than a true Cassegrain (Fig. 2-10). The Dall-Kirkham has extensive coma, and its usable field of view is smaller than that of the true Cassegrain.

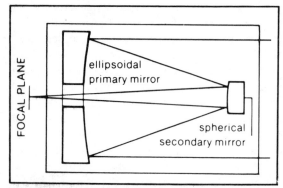

Fig. 2-10. The Dall-Kirkham has the disadvantage of extensive coma (courtesy Celestron International, Torrance, California, U.S.A.).

The Ritchey-Chretien reduces coma to almost zero, but it suffers from astigmatism and severe curvature of field (Fig. 2-11). Neither system is manufactured in large quantities. They can be very expensive compared to alternative catadioptric systems.

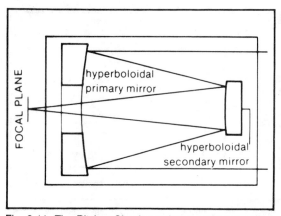

Fig. 2-11. The Ritchey-Chretien reduces coma but suffers from astigmatism and severe curvature of field (courtesy Celestron International, Torrance, California, U.S.A.).

CATADIOPTRICS

The catadioptric instruments combine the best ingredients of other telescopes. These instruments are relatively new and have been commercially available for only about 20 years. They are now the most popular instruments marketed in 3½-inch and larger apertures.

Catadioptrics have closed tubes that eliminate image-degrading tube currents and make the optics basically maintenance-free. The instruments are extremely portable due to the folding of the optical paths. The catadioptrics are free of chromatic aberration. The additional optical elements permit correction of various off-axis (or edge-of-field) aberrations (coma), especially with the Schmidt-Cassegrain and somewhat with the Maksutov. Catadioptrics are easy to set up and use. Large apertures still come in relatively small packages.

The catadioptrics cost more than Newtonian reflectors of equal aperture, but they are still considerably less expensive than refractors. Another disadvantage of most catadioptrics is that they have longer focal lengths (f/10 to f/20). The long ones (f/15 and longer) are good for lunar and planetary work. The f/10 systems are a good optimum ratio for lunar, planetary, and deep sky work. The Newtonians of fast speeds (f/5 or so) make the deep sky objects appear brighter than do the catadioptrics.

Many catadioptric instruments are ideal for terrestrial observing and/or photography, with near focus of 10 feet or less. Speeds of f/10 or faster are best. *Collimation* (alignment) is seldom a problem whether you are using an adjustable instrument or a permanently aligned Maksutov.

The catadioptric system is a combination mirror-and-lens system offering long focal length in a compact, closed tube. The lens is used ahead of the mirror to correct known faults in the mirror. It combines the best of both mirror-and-lens-only telescopes. There are two basic types of catadioptric systems—the *Maksutov-Cassegrain* and the

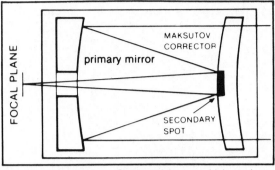

Fig. 2-12. The Maksutov-Cassegrain uses a thick meniscus correcting lens (courtesy Celestron International, Torrance, California, U.S.A.).

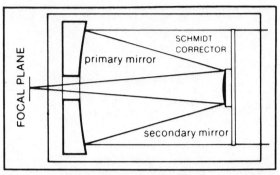

Fig. 2-13. The Schmidt-Cassegrain system uses a thinner aspheric correcting lens (courtesy Celestron International, Torrance, California, U.S.A.).

Schmidt-Cassegrain. The Maksutov uses a thick meniscus correcting lens (Fig. 2-12). The Schmidt system uses a thinner aspheric correcting lens (Fig. 2-13).

Chromatic aberration and coma are virtually nonexistent with the Schmidt system, but they are noticeable to a small degree with all Maksutov designs. Off-axis performance of the Schmidt is by far the best. The on-axis performance of all the folded systems mentioned is the same. If any one is pointed toward a point source at infinity, all the light entering the system will pass through a point at the focal plane of the lens. It is in off-axis performance that these telescopes markedly differ. The Schmidt-Cassegrain produces the sharpest images across the entire focal plane.

Maximum image contrast should be achieved for viewing certain low-contrast, narrow-field planets such as Jupiter and Saturn. The viewing condition (or air turbulence) is the single factor that most adversely affects image contrast when seeking planetary detail through a telescope. Instrument problems that can also adversely affect contrast are optical smoothness, baffling, and a small increase in central obstruction.

In a telescope that has no central obstruction and is otherwise perfect, 84 percent of the light energy from a point source is imaged in the primary maximum or Airy disk. Half of the remainder (7.2 percent) is in the first diffraction ring. Because the area of this first ring at the image plane is approximately six times greater than that of the Airy disk, the actual apparent intensity of this first ring will be 1.8 percent of the primary maximum's intensity. If the energy in the primary maximum is reduced by 10 percent with the introduction of a central obstruction, the energy in the first ring would increase from 7.2 percent to 16 percent. The visual intensity of the first ring would be increased relative to the primary maximum from 1.8 percent to 6 percent—a change that would hardly be detectable. The relative intensity of the other diffraction rings would be so low as not to even warrant consideration.

On a night of good viewing, use the Cassini division of Saturn's rings as a test subject. Vary the size of the central obstruction by using a set of cardboard circular masks of various diameters. See if you can tell the difference in contrast with the various masks. A very gross change in central obstruction would be required before any noticeable difference in contrast could be detected.

EYEPIECES

The *eyepiece* of a telescope magnifies the image formed by the objective lens or mirror. The magnifying power is determined by the ratio of the focal length of the telescope's objective to that of the eyepiece. The eyepiece design affects magnification and image brightness, flatness of field, chromatic and other aberrations, field of view, exit pupil, and eye relief. The eyepiece designer always faces the dilemma of minimizing one advantage in an effort to optimize another.

The Cross Optics company offers many quality eyepieces. The effective exit pupil is that of the eyepiece which results from the magnification-clear objective aperture combination and is independent of the eyepiece design. The eyepiece's actual exit pupil should be larger than the effective exit pupil, or the telescope's clear aperture will be effectively diaphragmed down from the eyepiece end. The limit as to how large the effective exit pupil can be is set by the observer's own pupils and is usually accepted to be 7 mm (in diameter). Some people have pupils that expand to more than 8 mm. It is best if the effective exit pupil is smaller than the eye's pupil or there will be no possibility of movement by the observer without vignetting. An exit pupil of about 5 mm is useful for a very wide field day/night low-power eyepiece. An effective exit pupil of 0.90-0.80 mm is required to make visible detail at the telescope's diffraction limit under optimum illumination-contrast conditions. Still smaller effective exit pupils (higher magnifications) are required for observing "limit" detail in many subjects.

Longer focal length eyepieces have greater

Table 2-1. Eyepiece Selection Guide (courtesy Cross Optics).

Magnification Range	Application	E*EP mm	f/4 FL	f/5 FL	f/6 FL	f/7 FL	f/8 FL	f/10 FL	f/11 FL	f/15 FL	f/16 FL	f/20 FL	f/25 FL
VERY LOW POWER	Richest (star) Field Telescope (RFT) observations; the most stars per view. Widest usable unvignetted field, assuming the observer's pupil expands to at least 7 mm	7.*	(*6.2) 25	35	(6.7) 40	(6.4) 45	(6.9) 55	(6.5) 65	(6.8) 75	—	—	—	—
	*Unless indicated in parentheses, figures are for EEP (*).												
LOW POWER	Wide field for viewing star fields, Earthshine, nebulae as well as daytime terrestrial viewing.	5.	20	25	30	35	40	45 (4.5)	55	75	75 (4.7)	75 (3.8)	—
		3.	12	16 (3.2)	20 (3.3)	20 (2.9)	25 (3.1)	30	35 (3.2)	45	55 (3.4)	55 (2.8)	75
MED. POWER	Studying variable stars, large area viewing of lunar features, viewing large nebulosities.	2.	8	10	12	16 (2.3)	16	20	25 (2.3)	30	35 (2.2)	40	55 (2.2)
		1.5	6	8	10 (1.6)	10 (1.7)	12 (1.4)	16 (1.6)	16 (1.5)	20	25 (1.6)	30	35 (1.4)
HIGH POWER	General use magnification normally most effective for planets, Moon, and study of globular star-clusters (.58 EEP excellent for Mars under optimal conditions).	1.0	4	5	6	6 (.86)	8	10	10 (.91)	16 (1.1)	16 (.90)	20	25
		75	3	4 (.80)	5 (.83)	5 (.71)	6	8 (.80)	8 (.73)	12 (.80)	12	16 (.80)	20 (.80)
		.58	5 (.63) +2 × B	3 (.60)	4 (.60)	4 (.67)	5 (.57)	6 (.60)	6 (.55)	8 (.54)	10 (.63)	12 (.60)	16 (.64)
VERY HIGH POWER	Precise collimating of telescope optics-observing close double stars, certain planetary details under perfect conditions, detail on satellites (Ganymede etc), viewing the Airy disk.	.50	4 +2 × B	5 +2 × B	3	3 (.43)	4	5	5 (.46)	6 (.40)	8	10	12 (.48)
		.38	3 +2 × B	4 +2 × B	5 (.42) +2 × B	5 (.31) +2 × B	3	4 (.40)	4 (.36)	5 (.33)	6	8 (.40)	10 (.40)

* Effective Exit Pupil

eye relief and allow more latitude in the placement of the eye near the eyepiece. Slower systems permit using long focal length eyepieces for high magnifications (effective exit pupils). Slower systems (f/15, f/20, etc.) permit more comfort in using higher magnifications. Unfortunately, these same instruments do not have the capability of the RFT (Rich-Field Telescope) mode with any existing eyepieces. A useful compromise may be achieved by using a Barlow lens, which can effectively turn a moderately fast system (f/6 to f/8) into a medium to slow system (f/12 to f/16 or f/14 to f/19) depending on whether the 2 × or 2.34 × Barlow lens is used.

A poor eyepiece can cause loss of definition and contrast on-axis and across the field, false color, distortion in extended objects, and asymmetric or expanded star images, with the resultant loss in effective light-gathering power. Table 2-1 is an eyepiece selection guide that may be helpful in providing the properties of the numerous eyepieces available.

☐ **Symmetrical**—Two identical achromats spaced a small distance apart.

☐ **Concave Lens**—Used in Galilean telescopes. This lens gives an erect image, but it provides only extremely low power and a very narrow field of view.

☐ **Ramsden**—Two plano-convex or double-convex lenses. This type is quite economical. The real image is formed in the front of the field lens, and the eyepiece has a 35 to 45-degree apparent field.

☐ **Orthoscopic**—Utilizes a triplet field lens and plano-convex eye lens. It provides excellent color correction and a wide 35 to 50-degree apparent field. This lens is found on more expensive telescopes.

☐ **Huygenian**—Consists of two simple lenses. The image plane falls between the two lenses. This system provides better correction than the Ramsden and has a 25 to 40-degree apparent field. The

Huygenian system is commonly found on both microscopes and telescopes.

☐ **Kellner**—This system provides good color correction and is excellent for refractor and longer focus instruments, with a 40 to 55-degree apparent field.

☐ **Erfle**—This design was formulated especially for extra-wide fields with good correction of all aberrations. The system is composed of three

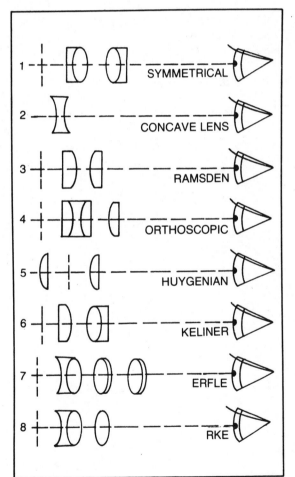

Fig. 2-14. Design configurations of various eyepieces (courtesy Edmund Scientific Company).

17

achromats and has an apparent field of view of 65 to 70 degrees. This type of design is usually quite expensive.

☐ **RKE**—This system is offered exclusively by Edmund Scientific and boasts computer-aided optimization of the Type II Kellner. The RKE eyepieces provide bright, wide, and sharp fields of view with good aberrational corrections using high index glass. The apparent field of view is 45 degrees.

Figure 2-14 depicts the eight systems with the different lenses and various design configurations. The manufacturer of your telescope can provide information to help you select the appropriate eyepieces. You should start off with possibly three or four eyepieces if possible. To aid you in the proper selection of eyepieces, the following order is suggested:

—Low power, 3 to 6 mm effective exit pupil (EEP), 8 × to 4 × per inch of telescope aperture for fields larger than the moon, nebulae, and star fields.

—Medium power, 0.8 mm EEP, 32 × per inch of telescope aperture for comparison of stars, variable stars, optimum for most planets.

—Very high power, 0.5 mm EEP, 50 × per inch of telescope aperture for double stars, centering of optics, near maximum for planets.

Chapter 3

Commercial Telescopes

M ANY COMPANIES MANUFACTURE TELE-
scopes and related astronomy accessories.
Your own viewing habits will aid you in making a
selection.

You will probably be tempted to go with one of
the large aperture models. These types gather
more light and will provide better celestial viewing.
If you need an instrument that can be easily packed
into the trunk of an automobile and carried to different sites, an overly large telescope will not be
practical. If you intend to permanently mount your
telescope at a home observatory, a larger model
may be ideal.

A Newtonian telescope is simple in design and
very efficient, but it can be impractical for certain
applications due to its length. If you want a large
aperture instrument but cannot tolerate the size of a
Newtonian design, you might opt for a Cassegrainian model, which offers the same aperture in a much
more compact package. This latter design is more
complex and expensive than the Newtonian.

SELECTING A TELESCOPE

There are many telescopes on today's market
that are more like toys than fine instruments. These
telescopes are often not professionally aligned, are
poorly made, and generally won't provide satisfactory performance for the serious astronomer. Be
wary of advertisements that place great importance
on the viewing power of a specific telescope but
give few additional specifications. The power of a
telescope is not nearly as important as its light-
gathering properties. High power is to be avoided
whenever possible.

Check the lens and mirror specifications, if
applicable. Some of the "toy" telescopes use cheap
plastic lenses that are optically unsuitable for any
serious celestial viewing. Even when optical spec-

ifications compare favorably from manufacturer to manufacturer, the physical mounting and construction specifications are of equal importance. A telescope that is not mechanically sound can be next to useless for the serious amateur. What materials are used for the tube assembly and mounting tripod? Is an equatorial wedge and motor drive included in the price? Does the manufacturer offer a comprehensive warranty with the instrument?

You need not spend thousands of dollars for a suitable instrument. My first telescope was manufactured by a children's scientific company that also offered many inexpensive chemistry sets, microscopes, and model rockets. The price of this 3-inch, 80-power Newtonian reflector was $17.95. A hollow aluminum tube was mounted atop the fiberboard main tube assembly. The telescope was mounted to a rather wobbly aluminum tripod. The telescope was focused by manually pushing the lens assembly into a plastic mounting platform located by setscrews to the fiberboard tube. The telescope was optically superb for its price. Many enjoyable hours were spent viewing the moon's craters, Saturn's rings, and the colorful bands on Jupiter's surface.

I gave this instrument to my son on his tenth birthday. When this telescope sustained wear and tear, I purchased another telescope for my son from the same local outlet as the original. A scientific company specializing in products for children manufactured this telescope. The telescope had a mechanically geared lens focusing system and the ability to align the main mirror. The price of the new telescope was more than $50; the other telescope cost less than $20. A fiberglass fiberboard replaced the hollow aluminum tube of the older model. The tripod was higher and much sturdier than that of the other telescope.

When I returned home from a business trip, my son was using the older telescope. The new instrument was completely unusable. Although perfectly aligned, cheap plastic lenses and a questionably ground mirror led to extremely poor light-gathering quality. Portions of the plastic lens mounting assembly had apparently melted during manufacture and occluded the viewing path. I rebuilt the new telescope with the optical components of the older model and produced an optically sound instrument with reasonably good mechanical stability.

Be wary of the printed specifications provided by some manufacturers who specialize in cheap rather than quality astronomical instruments. Stick with a manufacturer whose name you can trust and who will back up a product with a warranty. The manufacturers discussed in this chapter have had a long reputation for pleasing customers with their products.

CELESTRON INTERNATIONAL TELESCOPES

Celestron International is one of the best-known manufacturers of professional-quality telescopes, telephoto lenses, and binoculars. Celestron telescopes feature large optics. They employ lenses and mirrors to optically fold long high-power focal lengths into a compact configuration, resulting in observatory-size optics in a portable package. Because these telescopes are of large aperture, images will be bright and detailed. Clear, detailed images will be produced even at high power.

By day, a Celestron telescope can be used for long-distance macroscopy, nature studies, sports action, and candids. You can easily bring into sharp focus the antennae of a butterfly at 15 feet or the facial features of a person at a half-mile distance. At night under adverse city lighting conditions, you can study the moon and planets from your backyard. You can observe Venus's moonlike phases, Mars's surface features, Jupiter's cloud belts, or Saturn's rings. You can view the star clouds of the Milky Way, galactic star clusters, and planetary nebulae under dark skies.

Celestron telescopes are simple to set up and use. They feature sturdy fork mounts, slow-motion controls for fine adjustments in telescope point, and instant-lock clamps. The eyepiece and observing controls are conveniently located within inches of each other. The compact tube and fork mount rapidly dampen wind and mechanical vibrations, assuring image stability for astronomical observing or guided deep sky exposures. A closed tube design eliminates image-degrading air currents inside the tube. It also seals the tube against dust and other contaminants, assuring years of trouble-free service.

Finely etched star-locating circles make it easy to dial celestial objects into the field of view when the telescope is equatorially mounted. In the base of the telescope a system of motors and gears compensates for the earth's rotation during astronomical observations. This electric clock drive keeps celestial objects centered automatically in the field of view.

Many accessories provide increased capabilities. By adding other accessories to one of Celestron's telescopes, you can expand it into a research instrument or even into an astrophotographic laboratory.

Celestron 5

The Celestron 5 offers portability at an aperture large enough for both serious astronomical observing and photography. This 300-power telescope weighs only 13 pounds and is as portable as a briefcase (Fig. 3-1). The instrument features a daytime telescope, a 25 × telephoto lens, an astronomical tabletop observatory, and a deep space camera. This telescope is a good choice for the beginning amateur astronomer and nature enthusiast.

The Celestron 5 provides excellent viewing for daylight nature studies. You can observe the finest details at distances of 500 feet or more.

You can view the moon with all its crater formations, mountain chains, valleys, or riverlike rills in detail. The other planets in the solar system can

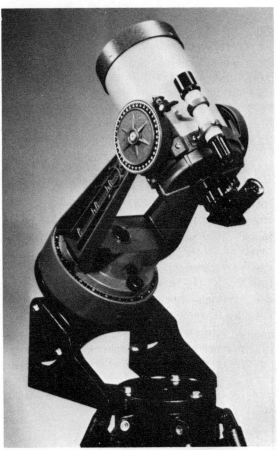

Fig. 3-1. The Celestron 5 Schmidt-Cassegrain telescope (courtesy Celestron International, Torrance, California, U.S.A.).

also be observed. Open clusters like the Pleiades or M35, globular clusters such as M13 or M22, emission nebulae like the Orion nebulae or the Veil nebula, planetary nebulae such as the Dumbbell nebula or the Ring nebula, and galaxies like the Whirlpool galaxy or M33 can be seen with the Celestron 5.

Convenient adapters are available as options that attach to the Celestron 5 and permit telephotography, long-distance macrophotography, full-disk lunar or solar photography, close-up lunar or solar photography, planetary photography, deep sky photography, and constellation photography.

The 5-inch aperture of the Celestron 5 may seem too small to be used as a deep sky camera. This small instrument may be fully equipped for deep sky guided astrophotography. The small aperture of the instrument generally spells "long exposure time" with tedious guiding when going after faint deep sky objects. This is offset by some special accessories that can be used for the Celestron telescopes. The cold camera offered as an accessory allows a reduction of exposure time by a factor of 10 for certain films. The use of the Tele-compressor reduces the exposure time by a factor of four. This combination allows superb quality deep sky color photographs. Wide fields swaths of the Milky Way may be captured on film using your 35-mm camera with its normal 50-mm taking lens. It is mounted on the Celestron 5 with the optional piggyback camera mount. The telescope serves as a stable platform during the exposure.

Table 3-1 provides complete information on the Celestron 5 regarding pertinent characteristics, features, and accessories available from the manufacturer. The compound catadioptric system of the Celestron 5 uses a combination of mirrors and lenses in the basic objective system. By optically folding the light path within the short tube, you have the performance of a 4-foot-long refractor with none of the image-degrading tube currents. The mirror system assures absolute freedom from any color distortions, but the closed tube assures freedom from contamination of the delicate first-surface mirrors, thus assuring years of maintenance-free operation.

Celestron 8

The Celestron 8 is portable, easy to use, and

Table 3-1. Specifications and Other Information for the Celestron 5 (courtesy Celestron International, Torrance, California, U.S.A.).

Characteristics, Features, and Accessories	
Optical configuration	
Schmidt-Cassegrain	
Clear Aperture .... 5″ (125 mm)	
Effective Focal Length .... 50″ (1,250 mm)	
Photographic Speed (focal ratio) f/10	
Highest useful visual	
magnification .... 300 X	
Oculars (Eyepieces) included,	
barrel dia .... 96″ (24.5 mm)	
Focal lengths—25mm (50 power),	
12mm (100X)	
Visual Back (Eyepiece tube)	
Diameter .... 96″ (24.5 mm)	
Right Angle Viewing—	
Star Diagonal .... 96″ Prism type	
Closest Focus .... Approx. 15 feet	
Resolution—(Dawes Limit)—	
arc-seconds 0.9	
lines per millimeter 197	
Stellar Magnitude Limit 13	
Image Scale—degrees per inch .... 1.15°	
Film Coverage (35mm) at	
30 feet .... 7″×10″	

100 feet .... 23″×35″	
1,000 feet .... 19′×29′	
Photographic Accessories .... Optional	
Mounting (equatorial wedge	
optional) .... Fork Type	
Slow Motion Controls, both	
axis .... Manual	
Finder Scope .... 5 power 24 mm	
Electric Clock Drive, dual synchronous	
motors .... 110v, 60Hz. 6 watts	
Setting Circles	
Declination—4″, R.A. 8″	
Drive Gear Diameter .... 6″ spur	
Polar Axis Diameter .... 1⅜″ tapered	
Tube Dimensions	
5½″ Dia. × 11″ Long	
Secondary Obstruction .... 16%, 2″ Dia.	
Tube Weight .... 4¼#	
Telescope Weight .... 13#	
Size, swung down .... 8″×9″×18″	
Carrying Case	
Dimensions .... 9″×12″×24″	
Shipping Weight .... 28#	

versatile (Fig. 3-2). Although it weighs only 21 pounds and takes up less room than a suitcase, this 480 × telescope delivers good optical performance to both the serious amateur and the research scientist. It can also double as a daytime telescope and 40 × telephoto lens.

You can view the textures of everyday objects in complete clarity with the Celestron 8. Sunspot structure, the solar transits of Mercury or Venus, lunar craters within lunar craters, Jupiter's and Saturn's cloud belts, and the surface features of Mars can be observed. Through the C8 under good seeing conditions, the cloud belts of Jupiter appear discontinuous and are composed of a multitude of streamers and festoons. At this aperture the oranges, reds, and browns of the Jovian clouds are revealed. The moons of Jupiter appear as disks and can be followed across the planet's entire face.

Saturn's rings are very detailed with the Celestron 8. The main division (Cassini's division) is instantly obvious even at low power. The innermost (crepe) ring has a distinctly rosy appearance even against the background of space. There is a noticeable brightening of inner ring B and pronounced darkening of outer ring A. Two cloud belts and six of Saturn's moons are visible through the Celestron 8. Even Mars (with about half the diameter of Earth) is surprisingly detailed, especially during favorable oppositions. Such surface features as Syrtis Major and Mare Erythraeum appear discontinuous. The clouds of the planet are visible, and the facing polar cap and its melt band can be seen.

Under dark skies the Celestron 8 brings its light-gathering power and resolution to bear for the deep sky observer. With more than 800 times the light-gathering power of the human eye, this telescope presents star clusters, nebulae, and even galazies in intricate detail. At this aperture the translucence that characterizes the planetary nebulae begins to emerge, and globular clusters are resolved to the core.

Open clusters such as the Pielades or M35, globular clusters like M13 or M3, and emission nebulae such as the Orion nebula can be seen with the Celestron 8. Through the Celestron 8:

☐ The nebulosity surrounding the stars of the Pieiades and what to the unaided eye appears to be six or seven stars become hundreds of stars.

☐ The open cluster of M35 reveals its fainter members, presenting a jewel boxlike appearance. Its companion cluster is strikingly noticeable.

☐ Globular cluster M13 is resolved to the core, and its central background blaze is dramatically bright.

☐ The tiny globular M3 is resolved to the core.

Fig. 3-2. The Celestron 8 Schmidt-Cassegrain telescope (courtesy Celestron International, Torrance, California, U.S.A.).

□ The Orion nebula glows with a multitude of intricately detailed filaments, and the four stars of the Trapezium become six or more stars.

□ The dark lanes of the Trifid nebula are obvious, and the trisection of the brighter component begins to become evident.

□ The fainter outer regions of the Dumbbell nebula become a network of delicate contrast levels.

□ The Ring nebula blazes forth brightly and distinctly.

□ The Whirlpool galaxy reveals a hint of spiral structure directed toward its companion galaxy.

□ The galaxy M33 reveals faint knots of stellar associations and its giant nebulosity MGC 604.

With the addition of optional adapters and attachments, the Celestron 8 can be used for many photographic applications. This telescope is entirely capable of deep sky photography with accessories. Table 3-2 provides complete specifications for the Celestron 8.

Celestron 11

The Celestron 11 (C11), Celestron's newest telescope, is for those who require a much larger light grasp and focal length in a portable package than is possible with a conventional telescope. The Celestron 11 features huge 11-inch clear aperture Schmidt Cassegrain optics of 2800 mm focal length at f/10. Despite its large aperture, the C11, with complete drive system, fork mount, and tube assembly, weighs only 57 pounds. Its tube assembly is less than 2 feet long. You can easily dismount and transport the Celestron 11 to remote sites.

The Celestron 11 is useful in astrophotography, nature studies, super telephoto photography, surveillance, and many other applications. Large aperture, portability, an extensive array of accessory options, sand-cast fork mount, and an accurate worm gear drive make it a versatile and capable product. Table 3-3 lists specifications.

Table 3-2. Specifications and Other Information for the Celestron 8 (courtesy Celestron International, Torrance, California, U.S.A.).

Optical configuration	Schmidt-Cassegrain
Clear Aperture	8″ (200 mm)
Effective Focal Length	80″ (2,000 mm)
Photographic Speed (focal ratio)	f/10
Highest useful visual magnification	480 X
Oculars (Eyepieces) included, barrel dia.	1¼″
Focal lengths—25mm (80 power), 40mm (50X)	
Visual Back (Eyepiece tube) diameter	1¼″
Right Angle Viewing— Star Diagonal (Prism type)	1¼″
Closest Focus (approx.)	25 feet
Resolution—arc-seconds (Dawes Limit)	0.6
lines per millimeter	210
Stellar Magnitude Limit	14
Image Scale—degrees per inch	0.72°
Film Coverage (35mm) at	
30 feet	4″×6″
100 feet	14″×21″
1,000 feet	11′×17′
Photographic Accessories	Optional
Mounting (equatorial wedge optional)	Fork Type
Slow Motion Controls, both axis	Manual
Finder Scope	6 power 30 mm
Electric Clock Drive, dual synchronous motors	110v, 60Hz. 6 watts
Setting Circles	Dec.—4″, R.A. 8″
Drive Gear Diameter	6″ spur
Polar Axis Diameter	1⅜″ tapered
Tube Dimensions	9″ Dia.×17″ Long
Secondary Obstruction	12%, 2¾″ Dia.
Tube Weight	11¼#
Telescope Weight	21#
Size, swung down	9″×13″×24″
Carrying Case Dimensions	13″×16″×30″
Shipping Weight	43#

Table 3-3. Specifications and Other Information for the Celestron 11 (courtesy Celestron International, Torrance, California, U.S.A.).

Standard features in the base price:

The base price Celestron 11 is a complete working telescope and includes: Optical Tube Assembly, Visual Back - 1¼ ", Star Diagonal - 1¼ ", 10 × 40 Finderscope, Fork Mount, Setting Circles, Electric Clock Drive, Clock Drive Cord, Piggyback Mount, Tele-Extender, T-Adapter, Counterweight Bar Assembly, 40mm - 1¼ " Ocular, 25mm - 1¼ " Ocular, 12mm - 1¼ " Ocular, Lens Cap, Carrying Cases, and Instruction Manual. Special Coatings, Wedge and Tripod are optional. (For other options see Celestron List Prices sheet.)

Clear Aperture: 11 inches (280mm)
Optical design: Schmidt-Cassegrain catadioptic; diffraction limited
Focal ratio: f/10
Focal length: 110 inches (2800mm)
Light grasp (compared to eye with 7mm pupil size): 1778x
Highest useful magnification: 660 ×
Lowest useful magnification: 42 × (approx. a 65mm ocular)
Photographic resolution: 200 lines/mm (theoretical at 4100 angstroms)
Standard oculars (eyepieces): 40mm - 1¼ ", 70 power; 25mm - 1¼ " ocular, 112 power; 12mm - 1¼ " ocular, 233 power.
Closest focus: 70 feet
Resolution (Dawes limit): 0.4 arc seconds
Stellar magnitude limit: (visual; approx.) 14.5
Image scale: 0.52 degrees/inch (= 0.02 degrees/mm)
Photographic field of view: 0.73 × 0.46 degrees on a 35mm slide format
Primary mirror:
  figure: spherical
  diameter: 11.2" (284.5mm)
  f/ratio: f/2.02
  radius of curvature: 44.5 inches (1131.6mm)
  material: fine annealed Pyrex
Secondary mirror:
  figure: spherical (final hand figuring yields a slightly aspheric figure)
  diameter: 3.1" (78.7mm)
  radius of curvature: 12.8" (325.1mm)
  material: fine annealed Pyrex
  amplification ratio: 4.95
Central obstruction: 4" (13.2% by area or 36.4% by diameter)
Corrector plate: aspheric Schmidt curve on exterior; plane interior
Thickness of corrector: ³⁄₁₆"
Back focus: 9 inches approx. maximum (from apex of primary)
Mirror coatings: enhanced aluminum; with silicone monoxide (SIO) protective overcoat
Optional special coatings (on both sides of corrector plate only): anti-reflection magnesium fluoride (MgF₂); 1/4 wave thickness optimized for 540NM.
Mount type: fork
Fork dimensions (max.):
  height: 26.5"

width: 18.3"
height of fork arm: 19¼ " (from top of drive base)
distance from center of declination axis to drive base: 16¾ "
distance from bottom of drive base to center of dec axis: 23³⁄₈"
width of fork arm: 3¾ "
Weight of fork mount and drive: 29.5 lbs.
Fork mount and drive castings: sand cast aluminum; type 356-T6
Optical tube assembly dimensions (max.):
  width: 12.5"
  length: 23.5"
  weight: 27.5 lbs.
Tube construction: aluminum
Drive system: worm gear; solar drive rate
Diameter of worm wheel: 6⅞"
Number of teeth on worm wheel: 216 (32 pitch, 14.5° pressure angle, pitch diameter 6.750 ± .001)
Power requirements: 110 volts, 60 Hz, 2 watts (other voltages and frequencies available)
Slow motion controls: manual standard on both axes (optional electric in declination)
Declination slow motion rate: 0.12°/revolution of control knob
R.A. slow motion rate: 1.8°/revolution of control knob
Polar axis: tapered
Diameter of north ball bearing: 1⅜"
Diameter of south ball bearing: 2¼"
Diameter of declination setting circle: 5⅛" (1° divisions)
Diameter of R.A. setting circle: 8¾" (5 min.-1.25° divisions)
Fasteners: stainless steel
Color: orange and brown (3 coat, 5-step process - primary coat, flat coat, bake, spatter coat, bake)
Interior color: flat black baked enamel and flat black aluminized
Lens cap: black anodized spun aluminum
Standard finderscope: 10 × 40 (8 × 50 and 10 × 70 optional accessories)
Filter provisions: 1. Photographic: optional standard Series 6 drop-in; install between T-adapter and rear cell of telescope. 2. Visual: Celestron eyepieces accept our optional eyepiece filters (.96" 1¼", 2" sizes available).
Rear cell threads: (Same as on rear cell of C5, C8, C14 reducer plate.) The C11 rear cell incorporates a removable C14 reducer plate for attachment of accessories via Universal Celestron Threads, 24 pitch thread, 2-inch diameter; rear cell baffle-tube-lock-nut threads (what the reducer plate threads onto), 16 pitch thd./3.290" diameter
Internal bore diameter of reducer plate: 1.5"
Internal bore diameter of main baffle tube: 2⅛"
Weight of optional wedge: 26 lbs.
Latitude adjustment range: 25°-68° (without latitude adjuster)
Weight of optional tripod: 33 lbs.
Size of wedge's shipping container: 10 × 14 × 23"
Size of tripod's shipping container: 49 × 13 × 12"
Notes:
1. The C11 utilizes the same tripod as the C14.
2. The C11 is shipped in four cases: optical tube assembly, fork mount/drive and accessories, optional tripod, optional wedge.
3. All specifications are approximate and Celestron International reserves the right to revise the instrument, accessories and specifications without notice.

Celestron 14

The Celestron 14 is a portable telescope with an aperture approaching professional size (Fig. 3-3). The company states that in 10 minutes you can demount the Celestron 14, load it into a compact car, and be on your way. This telescope can be installed on a pier under a permanent dome. The Celestron 14 is the ideal instrument for the advanced amateur astronomer, university observatory, or science center. Features include an aperture suitable for advanced research projects or student training, special high-transmission optical

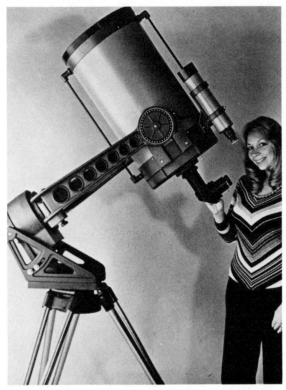

Fig. 3-3. The Celestron 14 Schmidt-Cassegrain telescope (courtesy Celestron International, Torrance, California, U.S.A.).

can even be glimpsed within the Red Spot and on the largest Jovian moon Ganymede (only when seeing conditions are most favorable).

The major divisions of Saturn's ring system will be immediately obvious, and you can notice Saturn's surface detail. There is a striking amount of banded detail on the globe, usually including one or more white spots, and a hint of belt structure near the polar region. Saturn's moon Titan appears as a disk. At this aperture your viewing possibilities include virtually the entire solar system—tiny lunar rills broken by even tinier craterlets, the bluish-green disks of Uranus and Neptune, and the pinpoint image of Pluto moving from month to month across faint star fields. At this aperture globular clusters are resolved so well that each seems to have its own personality. Galaxies are revealed in intricate detail, and even the brighter quasars may be reached. Through the Celestron 14:

☐ The tiny companion cluster of the open cluster of M35 is resolved.

☐ Open cluster M11 is revealed as a spectacular gathering of hundreds of stars.

☐ Numerous dark lanes and streamers of stars radiate outward from the cores of the globular clusters of M13 and M3.

☐ The intricate filamentary network of the Orion nebula is laced with many knotty brightenings, and the dark clouds of the nebula are crisply defined.

☐ Both components of the Trifid nebula are well-detailed, with the brighter one clearly trisected by dark lanes.

☐ The Dumbbell nebula becomes a complete oval with scores of stars apparently embedded in its gossamer beauty.

☐ The contrast levels of the Ring nebula are evident, and its central star may be glimpsed occasionally.

☐ The nucleus of the Whirlpool galaxy is very bright. Its spiral structure is quite distinct, with an arm connecting its companion galaxy.

coatings, stable fork mounting, and remote-controlled electrical slow motions in both axes.

Although the Celestron 14 weighs a little more than 100 pounds, its components weigh no more than 50 pounds each. Transport to a dark sky observing site is easier than with some smaller, conventionally mounted classical telescopes.

The Celestron 14, with 2,580 times the light-gathering power of the human eye, is a superior instrument for lunar, planetary, or deep sky observations. Through this telescope, when the air is steady, the cloud belts of Jupiter are delicately festooned and display an enormous range of colors—cream, orange, and gray—and within the belts are numerous smaller storms (white spots). Detail

Table 3-4. Specifications and Other Information for the Celestron 14 (courtesy Celestron International, Torrance, California, U.S.A.).

Characteristics, Features, and Accessories

Optical configuration .... Schmidt-Cassegrain
Clear Aperture .... 14" (350 mm)
Effective Focal Length .... 154" (3,900 mm)
Photographic Speed (focal ratio) f/11
Highest useful visual
    magnification .... 840 X
Oculars (Eyepieces) included,
    Barrel dia .... 1¼"
        40mm focal length—100X;
        25mm—160X
        12mm—325X; 6mm—650X
Star Diagonal (right angle
    viewing) .... 2" mirror type
    accepts 2" oculars and includes
    1¼" adaptor.
Closest Focus (approx.) .... 100'
Resolution (Dawes Limit)—
    arc seconds 0.3
    lines per millimeter 181
Stellar Magnitude Limit 15
Image Scale - degrees per inch 0.37
    Full Lunar Disc Diameter .... 1.3"
Mounting (equatorial wedge
    optional) .... Fork Type
Photographic accessories included in
base price
    T-Mount Camera Adaptor,
        Tele-Extender
    Piggyback Camera Mount,
        Counterweight Set
Finderscope .... 10 power, 40 mm
Electric Clock Drive—
    synchronous motor
Slow Motion Control—dual speed,
    both axes.
Power requirement ...... 110v, 60Hz., 10 watts
Setting Circles .... Declination - 6", R.A. 9½"
Drive Gear Diameter .... 6¾" worm
Polar Axis Diameter .... 3" tapered
Tube Dimensions .... 16" Dia., 30" long
Secondary Obstruction .... 10%, 4¼" Dia.
Tube Weight .... 50#
Telescope Weight (tube assy, mount,
    drive) .... 108#
Size, swung down .... 18"×22"×44"
Case Dimensions, Tube .... 21"×22"×36"
Mount and Accessory .... 12"×21"×36"
Shipping Weight .... 200#

□ Galaxy M82 resembles a curved needle, with many knotty brightenings toward its center.

Convenient adapters, some of which are included in the base price of the Celestron 14, provide a means of attaching a 35-mm camera body or body plus lens to the instrument. You can use the Celestron 14 for telephotography, wide-field lunar photography, close-up lunar and planetary photography, basic deep sky photography, and constellation photography. This telescope can be converted to a full-fledged astrophotographic research station with the right accessories. Table 3-4 lists specifications and features of the Celestron 14.

Celestron 90 Astro Telescope

The Celestron 90 (C90) is the most compact and lowest-priced Celestron telescope (Fig. 3-4). It

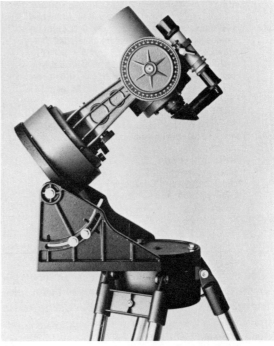

Fig. 3-4. The Celestron 90 Maksutov-Cassegrain telescope (courtesy Celestron International, Torrance, California, U.S.A.).

Table 3-5. Specifications and Other Information for the Celestron 90 (courtesy Celestron International, Torrance, California, U.S.A.).

```
Optics . . . . . . . . . . . . . . . . . Maksutov-Cassegrain
Aperture . . . . . . . . . . . . . . . . . . . . . . 90mm (3.55")
Focal Length . . . . . . . . . . . . . 1000mm (40 inches)
Focal Ratio. . . . . . . . . . . . . . . . . . . . . . . . . . f/11
Photographic Power . . . . . . . . . . . . . . . . . . . 20×
Resolution (Dawes limit) . . . . . . . 1.3 arc seconds
                                        (120 lines/mm)
Visual Stellar Magnitude Limit. . . . . . . . . . . . 12
Image Scale . . . . . . . . . . . . . . . . . . . . . 1.43"/inch
Near Focus (approx.) . . . . . . . . . . . . . . . 10 Feet
Secondary Obstruction. . . . . . . . . . . . . 1⅜"(15%)
Finderscope** . . . . . . . . . . . . . . . . . . 5 × -24mm
Slow-Motion Controls* . . . . . . Manual (both axes)
Electric Drive* . . . . . . . . . . . . . 3 watts, 110v 60Hz
                            (unless otherwise specified)
Drive Gear Diameter* . . . . . . . . . . . . 4½ inch spur
Setting Circle Diameter*  R.A. 6¼ inches (Driven)
                                   Dec.:  4 inches
Visual Back (eyepiece tube) . . . .96" O.D. (built-in)
Eyepiece (ocular)** . . . . . 18mm Kellner, .96" O.D.
Barlow Lens** . . . . . . . . . . . . . . . 2.5 × .96" O.D.
Star Diagonal** . . . . . . . . . . .96" O.D. Prism type
Telephoto/Spotting Scope Size 5" Dia. × 8" Long
   Weight. . . . . . . . . . . . . . . . . . . . . . . 3# (approx.)
Telephoto/Spotting Scope Case
   Size . . . . . . . . . . . . . . . . 6½" × 8½" × 10½"
   Shipping Weight . . . . . . . . . . . . . . 8# (approx.)
Telescope Size (swung down). . . . . 7" × 7" × 13"
   Weight. . . . . . . . . . . . . . . . . . . . . 8# (approx.)
Telescope Case Size . . . . . . . . 8½" × 8½" × 15"
   Shipping Weight . . . . . . . . . . . . . 17# (approx.)
   *Only on C90 Astro Telescope
   **Only on C90 Astro Telescope and Spotting
     Scope
```

is unusually small and lightweight for its focal length (1000 mm). This makes it useful as a telephoto lens or spotting scope. To assist in these applications, Celestron provides accessories designed around the C90 optical system.

The heart of the C90 series is the 90-mm aperture, f/11, Maksutov-Cassegrain optical system, which is identical on all three C90 versions. The C90 is really a telescope that is also a telephoto. By adding the appropriate optional accessories, it is possible to use any C90 version in another mode.

The C90 has advanced features like star locating setting circles, an automatic tracking system, and slow-motion controls on both axes. The standard ocular and Barlow lens provide magnifications of 55 ×, 140 × and 200 ×.

The C90 is quite powerful as a daytime telescope. You can easily recognize people at distances of one-half mile away or read an automobile's license plate two miles away. There is much to see in the night sky with the Celestron 90. The moon will display craters, rugged mountains, rills, and relatively smooth plains. You can see Jupiter's equatorial cloud belts and its great Red Spot. You can also watch the four Galilean satellites as they orbit Jupiter. Saturn's rings will be easily visible through this telescope. Deep sky objects such as the Great Nebula in Orion; the Andromeda galaxy; the Ring, Lagoon, Trifid and Dumbbell nebulae; the great Hercules cluster; and other objects can be seen when the skies are dark.

Adapters are available to allow the attachment of a camera and make this telescope even more versatile. The Celestron 90 is most suitable for the beginning amateur astronomer, nature enthusiast, or photo hobbyist who desires quality at a relatively low price. Table 3-5 provides complete specifications and dimensions on the Celestron 90.

EDMUND SCIENTIFIC COMPANY TELESCOPES

Edmund Scientific Company offers quality telescopes, laboratory and photography equipment, and learning toys for children. Edmund Scientific is basically a mail-order firm, but there is a retail store in New Jersey where you can try out their instruments. To obtain a catalog, write to Edmund Scientific Company, 101 E. Gloucester Pike, Barrington, NJ 08007.

Voyager 6001/6008 Series

Figure 3-5 shows the Voyager telescope, which is designed for both astronomical and terrestrial use. It features a "Rank" achromatic objective lense, which has been developed expressly for Edmund Scientific by Dr. David Rank, a renowned optical designer. Weighing only a scant 7 pounds, the Voyager is a truly portable instrument. The compact design makes it ideal for nature outings. The Voyager will provide sharp and detailed images of birds, animals, flowers, and insects.

The Voyager's optical system makes it possible to view lunar craters, Jupiter's satellites and rings, tinted double and multiple suns, star clusters, nebulae, and some brighter galaxies. Some of the Voyager's other features are:

☐ 15-mm RKE eyepiece—This ocular is supplied as standard equipment.
☐ Roof prism image erector—This provides fully corrected (left to right and right side up) images for both land and sky gazing.
☐ Telephoto lens—With accessory attachments, a 35-mm single lens reflex (SLR) camera can be coupled with the Voyager to film your land and sky targets.

Table 3-6 provides specifications and dimensions for this reasonably priced telescope.

Astroscan 2001

The Astroscan 2001 is a Newtonian rich-field reflector that is ideal for the beginning amateur

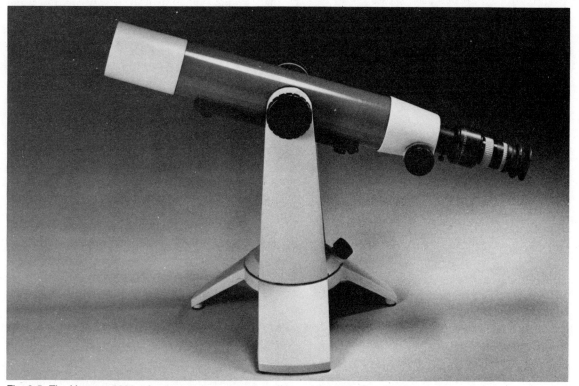

Fig. 3-5. The Voyager 6001 refractor telescope (courtesy Edmund Scientific Company).

Table 3-6. Complete Specifications for the Voyager (courtesy Edmund Scientific Company).

Specifications

Objective:	60mm dia. f/8 Rank achromat; 486 mm (19″) focal length; 1.9 arc second resolution; 100% inspected on a Zygo Interferometer
Eyepiece:	15mm e.f.l. RKE® 45° apparent field Standard 1¼″ OD barrel
Magnification:	32 X with 15mm RKE® To 150 X with optional eyepieces and Barlow
Field of View:	1.4° at 32 X
Image Erector:	Right-angle roof prism
Finderscope:	5× 24 mm cross-line recticle Eccentric ring mount
Mounting:	Table top altazimuth fork Aluminum alloy Rubber pads on base Convertible to standard camera tripod
Focuser:	Smooth rack and pinion Accepts standard (1¼″ OD) eyepieces and accessories
Coatings:	Anti-reflection throughout
Dimensions:	Height: 15½″ (38cm) Weight: under 7lb (3kg) Tube length: 18″ (44cm)

inition in order to increase magnification, wide-field telescopes accentuate the brightness and clarity of celestial objects through their remarkable light-gathering power. Where high-power telescopes magnify obscuring atmospheric disturbances along with the object viewed, wide-field telescopes remain relatively unaffected by earth's atmosphere. Even on occasions when viewing conditions are actually bad, you can sometimes glimpse stars through a wide-field telescope when none are visible to the unaided eye. Wide-field telescopes are perfect for scanning broad expanses of the cosmos.

The Astroscan 2001 is a Newtonian reflector telescope engineered inside a sphere. It utilizes a 4½-inch (108-mm) diameter; f/4.4 effective focal ratio; parabolic, precision-ground primary mirror that is polished and figured to ⅛ wave; a ⅛ wave diagonal mirror; and a 28-mm focal length com-

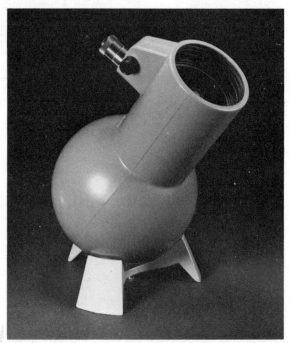

Fig. 3-6. The Astroscan 2001 Newtonian rich-field reflector (courtesy Edmund Scientific Company).

astronomer (Fig. 3-6). Its simplicity of design provides both portability and easy operation.

Wide-field telescopes, also known as rich-field telescopes or RFTs, show the most stars in one view. They provide you with spectacular views of the heavens by optimizing the balance between the angular field and light-gathering power. Figure 3-7 illustrates the field of view available with various instruments.

Unlike ordinary telescopes that rely on high magnification to resolve images of single celestial objects or small groups of stars, wide-field telescopes use relatively low magnification to obtain bright, clear images of star fields, comets, nebulae, galaxies, planets, and the moon. Where high-power telescopes sacrifice light intensity and image def-

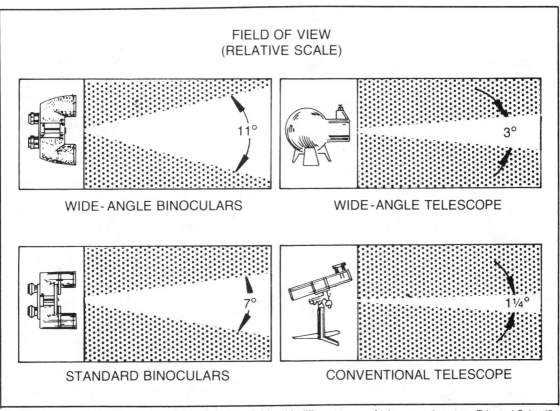

FIELD OF VIEW
(RELATIVE SCALE)

WIDE-ANGLE BINOCULARS 11°

WIDE-ANGLE TELESCOPE 3°

STANDARD BINOCULARS 7°

CONVENTIONAL TELESCOPE 1¼°

Fig. 3-7. A comparison of the various fields of view available with different types of telescopes (courtesy Edmund Scientific Company).

pound achromatic eyepiece with smooth, roller-type focusing to provide a broad, bright field of 3 degrees. This optical system is factory collimated for precise alignment and should never require adjustment under normal use. To enhance the telescope's minimum care performance, the diagonal mirror assembly is mounted directly to a coated optical window that provides protection against moisture, dust, and possible damage of the primary mirror.

Weighing only 10.4 pounds with its stand and only 17 inches long, the Astroscan 2001 is quite portable and easy to operate. Its lightweight, removable cast aluminum stand, shoulder carrying strap, and unusual spherical design enable you to take it anywhere.

Molded of high-impact plastic for rugged durability, the telescope body is an attractive red to prevent impairment of night vision. When placed in its stand, the advantages of the telescope's spherical design will become apparent. The bulk and complexity of traditional mounts are eliminated, but simple fingertip pressure is all that is required to orient the telescope to any desired angle.

With a magnification power of 16 × and an objective focal length of 17½ inches (445 mm), the Astroscan 2001 minimizes the image distortion caused by high magnification of atmospheric inter-

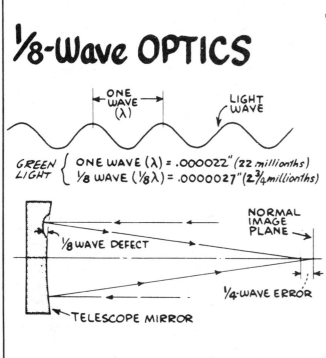

⅛-Wave OPTICS

ONE WAVE (λ)

LIGHT WAVE

GREEN LIGHT { ONE WAVE (λ) = .000022" (22 millionths)
⅛ WAVE (⅛λ) = .0000027" (2¾ millionths)

NORMAL IMAGE PLANE

⅛ WAVE DEFECT

¼-WAVE ERROR

TELESCOPE MIRROR

200 WAVES! —
SHEET OF PAPER

The usual tolerance for high-precision optics is one-quarter of the wavelength of light—no part of the glass surface must depart more than ¼ wave or 5½ millionths-of-an-inch from the specified shape. Compare this with a sheet of paper which has a thickness of about 200 waves! The Edmund Astroscan 2001 betters this tolerance with optics of 1/8 wave. In the case of a telescope mirror where light transverses the distance twice, a 1/8 wave defect on the mirror will result in a 1/4 wave error at the image plane as shown. This gives nearly perfect imagery. Further narrowing of the tolerance to 1/10 wave or less is more in the nature of advertising claims than any appreciable gain in definition.

Fig. 3-8. The Astroscan 2001 features ⅛ wave optics, which minimizes image distortion (courtesy Edmund Scientific Company).

ference. A bright and wide field of view is achieved at the same time.

The usual tolerance for high-precision optics is one-quarter of the wavelength of light. No part of the glass surface must depart more than ¼ wave or 5½ millionths of an inch from the specified shape. Compare this with a sheet of paper that has a thickness of about 200 waves. The Astroscan 2001 betters this tolerance with optics of ⅛ wave. In the case of a telescope mirror where light transverses the distance twice, a ⅛ wave defect on the mirror will result in a ¼ wave error at the image plane (Fig. 3-8). This gives nearly perfect imagery. Further narrowing of the tolerance to 1/10 wave or less is more in the nature of advertising claims than any appreciable gain in definition.

In the Astroscan 2001 parallel rays of light from a star or distant object enter the optical system through the anti-reflection coated optical glass window in the instrument's front part. The light rays then strike the parabolically curved primary mirror in the back of the telescope, where they are reflected and converge to a focal point after being reflected outside the telescope tube by the flat secondary mirror.

A miniature image of the object being viewed is formed at the focal plane. The eyepiece is then used to magnify the image in the same manner that a simple magnifying glass is used to magnify small print. The field stop in the eyepiece provides a sharp, circular boundary to the field of view. By placing the eye a short distance (about 1 inch) be-

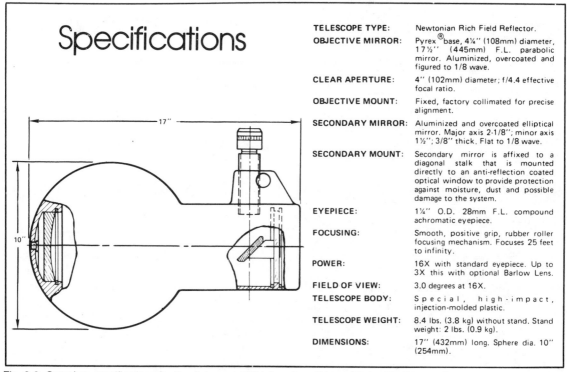

Specifications

TELESCOPE TYPE:	Newtonian Rich Field Reflector.
OBJECTIVE MIRROR:	Pyrex® base, 4¼" (108mm) diameter, 17½" (445mm) F.L. parabolic mirror. Aluminized, overcoated and figured to 1/8 wave.
CLEAR APERTURE:	4" (102mm) diameter; f/4.4 effective focal ratio.
OBJECTIVE MOUNT:	Fixed, factory collimated for precise alignment.
SECONDARY MIRROR:	Aluminized and overcoated elliptical mirror. Major axis 2-1/8"; minor axis 1½"; 3/8" thick. Flat to 1/8 wave.
SECONDARY MOUNT:	Secondary mirror is affixed to a diagonal stalk that is mounted directly to an anti-reflection coated optical window to provide protection against moisture, dust and possible damage to the system.
EYEPIECE:	1¼" O.D. 28mm F.L. compound achromatic eyepiece.
FOCUSING:	Smooth, positive grip, rubber roller focusing mechanism. Focuses 25 feet to infinity.
POWER:	16X with standard eyepiece. Up to 3X this with optional Barlow Lens.
FIELD OF VIEW:	3.0 degrees at 16X.
TELESCOPE BODY:	Special, high-impact, injection-molded plastic.
TELESCOPE WEIGHT:	8.4 lbs. (3.8 kg) without stand. Stand weight: 2 lbs. (0.9 kg).
DIMENSIONS:	17" (432mm) long. Sphere dia. 10" (254mm).

Fig. 3-9. Complete specifications for the Astroscan 2001 (courtesy Edmund Scientific Company).

hind the eyepiece, you see a sharp, clear, magnified image of the object being viewed.

The Astroscan 2001 is designed for optimum performance with magnifications of 15 × to 35 ×. It can be used with higher power eyepieces. This is not recommended though, because there will be a corresponding decrease in the field of view and a corresponding increase in difficulty in aiming the telescope at a particular object. Figure 3-9 provides complete specifications and dimensions for the Astroscan 2001.

Edmund 3-Inch Reflector Telescope

This 3-inch Newtonian reflector is an ideal first telescope for those new to astronomy who want to start out with a minimum investment (Fig. 3-10). This telescope is made with precision optical elements. The complete unit is pretested and collimated to make sure it is workable. You can see Saturn's rings, the Straight Wall on the moon, and many of the famous M-objects (including diffuse nebulae and open clusters of stars). In light grasp the telescope will show stars to 10½ magnitude, putting over 400,000 stars within reach of this telescope's big eye. Additional power up to 90 × can be obtained with accessory lenses. High power is used mainly for moon and planetary detail and splitting very close double stars. Table 3-7 contains specifications and pertinent dimensions for this telescope.

Edmund-Rank 4-Inch Astronomical Refractor

Figure 3-11 shows Edmund's 4-inch model,

Fig. 3-10. The Edmund 3-Inch Newtonian reflector (courtesy Edmund Scientific Company).

Table 3-7. Specifications and Other Information on the 3-Inch Reflector (courtesy Edmund Scientific Company).

- 3″ (76 mm) diameter f/10
  775 mm (30½″) f.l. spherical
  primary mirror, aluminized & overcoated:
  resolution 1.5 arc second
- 28|mm RKE® eyepiece for 28 K
- Adjustable 2 X to 3 X Barlow lens for 56 X to 84 X
- Standard size rack & pinion focuser
- Low expansion PVC telescope tube
- 5×24 mm finder with cross-hair reticle
- Cast aluminum fixed equatorial head with positive locking fork
- 36″ hardwood tripod with metal leg braces
- Field of view: 0.50° with 8 mm RKE®; 0.89° with 15mm RKE®; 1.75° with 28 mm RKE®
- Weight: 11.5 lbs (5.2 kg)

which is also available in a 3-inch configuration with the same design features. It is equipped with the American-made objective lenses designed by Dr. David Rank, which are crafted to exacting tolerances in Edmund's optical shops. Each lens is fully corrected for chromatic and spherical aberration and delivers clear and striking definition on high resolution targets such as the moon, planets, and close double stars. With the closed tube and minimum maintenance convenience of the classical refractor, both the 3-inch and 4-inch models will provide many years of enjoyable and pleasurable viewing.

Some features include:

☐ "Rank" objective, two-element air-spaced achromat coated on all surfaces that reduces unwanted reflections, maximizes light transmission—diffraction limited.

☐ Prism star diagonal—for right-angle viewing comfort of objects at high sky elevations.

☐ Smooth rack and pinion focusing—drawtube accepts standard size eyepieces and accessories.

☐ A 5×24-mm finder—with cross hair reticle and eccentric ring mount for easy alignment.

☐ Cast aluminum equatorial mount—with built-in setting circles, electric clock drive, and adjustable cradle strap for easy balancing. Nylon bearings and quick release knobs are on both axes.

☐ Camera adapters, image erectors, and a star spectroscope are available.

Table 3-8 shows pertinent specifications and other data on the 4-inch model. Table 3-9 lists information for the 3-inch telescope.

Edmund 4¼-Inch f/4 Telescope

The 4¼-inch model shown in Fig. 3-12 is the ideal Newtonian instrument for serious beginners and intermediate amateurs who desire the wide, bright views that only a rich-field telescope can offer. This telescope is built with the same preci-

Fig. 3-11. The Edmund-Rank 4-inch astronomical refractor (courtesy Edmund Scientific Company).

sion optical system as the Astroscan 2001. The 4¼-inch f/4 provides an unusually wide 3-degree field of view with the 28-mm RKE eyepiece supplied with the telescope. Also contributing to this wide view are the maximum resolution possible for the size of the instrument and good light-gathering power. The thermal insulating phenolic fiber tube is fitted with aluminum end rings. A

**Table 3-8. Specifications for the
4-Inch Refractor (courtesy Edmund Scientific Company).**

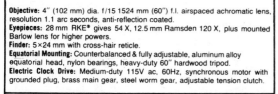

4-Inch f/15

Objective: 4″ (102 mm) dia. f/15 1524 mm (60″) f.l. airspaced achromatic lens, resolution 1.1 arc seconds, anti-reflection coated.
Eyepieces: 28 mm RKE® gives 54 X, 12.5 mm Ramsden 120 X, plus mounted Barlow lens for higher powers.
Finder: 5×24 mm with cross-hair reticle.
Equatorial Mounting: Counterbalanced & fully adjustable, aluminum alloy equatorial head, nylon bearings, heavy-duty 60″ hardwood tripod.
Electric Clock Drive: Medium-duty 115V ac, 60Hz, synchronous motor with grounded plug, brass main gear, steel worm gear, adjustable tension clutch.

**Table 3-9. Specifications for the 3-Inch Model
of This Refractor (courtesy Edmund Scientific Company).**

3-Inch f/11

Objective: 3″ (76 mm) dia. f/11 838 mm (33″) f.l. airspaced achromatic lens, resolution 1.5 arc seconds, anti-reflection coated.
Eyepieces: 28 mm RKE® gives 30 X, 12.5 mm Ramsden 67 X, plus mounted Barlow lens for higher powers.
Finder: 5 × 24 mm with cross-hair reticle.
Equatorial Mounting: Counterbalanced & fully adjustable, aluminum alloy equatorial head, nylon bearings, heavy-duty 60″ hardwood tripod.
Electric Clock Drive: Medium-duty 115V ac, 60Hz, synchronous motor with grounded plug, brass main gear, steel worm gear, adjustable tension clutch.

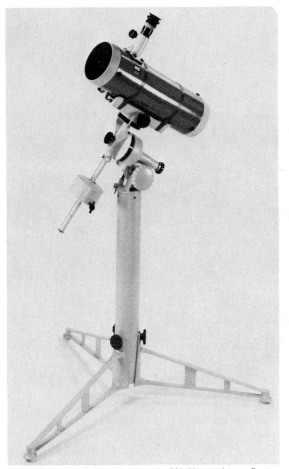

Fig. 3-12. The Edmund 4¼-inch f/4 Newtonian reflector (courtesy Edmund Scientific Company).

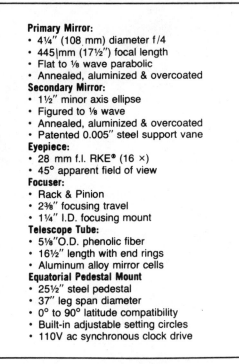

Primary Mirror:
- 4¼" (108 mm) diameter f/4
- 445|mm (17½") focal length
- Flat to ⅛ wave parabolic
- Annealed, aluminized & overcoated

Secondary Mirror:
- 1½" minor axis ellipse
- Figured to ⅛ wave
- Annealed, aluminized & overcoated
- Patented 0.005" steel support vane

Eyepiece:
- 28 mm f.l. RKE® (16 ×)
- 45° apparent field of view

Focuser:
- Rack & Pinion
- 2⅜" focusing travel
- 1¼" I.D. focusing mount

Telescope Tube:
- 5⅛"O.D. phenolic fiber
- 16½" length with end rings
- Aluminum alloy mirror cells

Equatorial Pedestal Mount
- 25½" steel pedestal
- 37" leg span diameter
- 0° to 90° latitude compatibility
- Built-in adjustable setting circles
- 110V ac synchronous clock drive

ity. Table 3-10 lists complete performance data and specifications for this versatile telescope.

Edmund 4¼-Inch f/10 Telescope

The 4¼-inch f/10 telescope is ideal for beginners and serious amateurs who want professional features at a low price (Fig. 3-13). The 4¼-inch aperture provides twice the light-gathering ability of 3-inch telescopes and sufficient magnification to explore deep sky objects. This telescope comes equipped with a sturdy, lightweight equatorial mount with synchronous clock drive and the 28-mm RKE eyepiece with rubber eyeguard. Made of rigid phenolic fiber tubing, the 4¼-inch f/10 provides excellent insulation. Fitted with cast aluminum end rings and patented super-thin (0.005-inch) ten-

patented, super-thin (0.005-inch) tensioned steel support vane for the secondary mirror reduces vibrations and provides unobstructed viewing. The lightweight equatorial pedestal mount is equipped with many professional features, including very strong fiber-reinforced polypropylene straps that firmly secure the telescope into its cradle, hysteresis synchronous clock drive, setting circles on both axes, 0 to 90-degree latitude compatibility, and quick-detach pedestal legs for increased portabil-

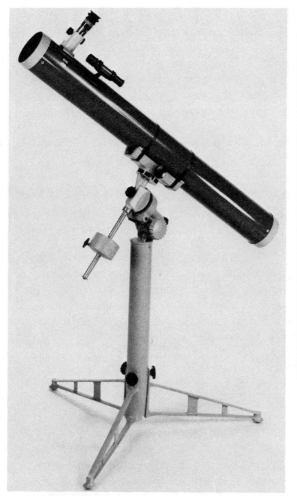

Fig. 3-13. The Edmund 4¼-inch f/10 telescope (courtesy Edmund Scientific Company).

Table 3-11. Specifications for the
4¼-Inch f/10 (courtesy Edmund Scientific Company).

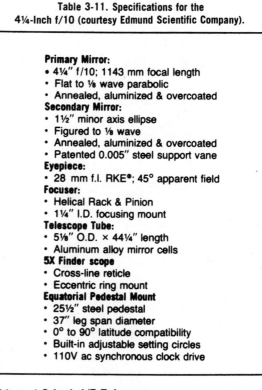

Primary Mirror:
- 4¼″ f/10; 1143 mm focal length
- Flat to ⅛ wave parabolic
- Annealed, aluminized & overcoated

Secondary Mirror:
- 1½″ minor axis ellipse
- Figured to ⅛ wave
- Annealed, aluminized & overcoated
- Patented 0.005″ steel support vane

Eyepiece:
- 28 mm f.l. RKE®; 45° apparent field

Focuser:
- Helical Rack & Pinion
- 1¼″ I.D. focusing mount

Telescope Tube:
- 5⅛″ O.D. × 44¼″ length
- Aluminum alloy mirror cells

5X Finder scope
- Cross-line reticle
- Eccentric ring mount

Equatorial Pedestal Mount
- 25½″ steel pedestal
- 37″ leg span diameter
- 0° to 90° latitude compatibility
- Built-in adjustable setting circles
- 110V ac synchronous clock drive

Edmund 8-Inch f/5 Telescope

Edmund's 8-inch f/5 telescope is a Newtonian system that offers wide views from rich field to diffraction-limited high resolution. It minimizes the optical aberrations common to compound optical systems, yet offers good light-gathering capabilities. This telescope will work well for prime focus astrophotography. The 8-inch f/5 telescope significantly reduces exposure times for deep sky objects, as compared to f/10 systems, and provides sharp full-frame exposures with minimum image distortion. The 2-inch inside diameter focusing mount provides an unrestricted view of the field.

The complete system weighs 70 pounds, making it easy to store and transport. The sturdy phenolic impregnated fiber tube provides thermal insulation and vibration damping along the optical

sioned steel vanes for rigid support, it provides stable and unobstructed viewing. Fiber-reinforced polypropylene straps firmly hold the telescope in the cradle of the completely redesigned equatorial mount. This telescope and its mount are remarkably easy to set up, operate, and transport to field locations. Table 3-11 contains all pertinent data and specifications on the 4¼-inch f/10 telescope.

path. The three-point adjustable primary mirror cell and fully adjustable secondary mirror cell, with a patented 0.005-inch super-thin rigid tension diagonal vane, maintain precise optical alignment and permit unobstructed viewing.

The lightweight equatorial fork mount offers stability, portability, and easy access to all areas of the sky. Low friction bearings provide smooth motion on right ascension and declination axes. This telescope is also equipped with setting circles on both axes, fine adjustment declination control, and a hysteresis synchronous motor. Many accessories are available for attachment and mounting to this versatile telescope, including an off-axis guiding system for prime focus astrophotography. Complete specifications on the 8-inch f/5 telescope are listed in Table 3-12.

Table 3-12. Specifications and Other Information for the 8-Inch f/5 (courtesy Edmund Scientific Company).

Primary Mirror:	• 8-Inch (203 mm) diameter f/5
	• 1016 mm focal length
	• Figured to ⅛ wave, parabolic
	• Annealed Pyrex, aluminized, & overcoated
Secondary Mirror:	• 2⅝″ minor axis ellipse
	• Flat to ⅛ wave
	• Annealed Pyrex, aluminized & overcoated
Eyepiece:	• 28 mm f.l. RKE® gives 36 ×
	• 45° apparent field of view
Finder:	• 9 × 30 mm achromatic objective
	• 6° field w/cross line reticles
	• Right-angle viewing comfort
	• Adjustable ring mounts
Focuser:	• 2″ I.D. Helical Rack & Pinion
	• 1¼″ I.D. eyepiece adapter
	• 2″ O.D. to "T" mount adapter
Telescope Tube:	• 9½″ O.D. phenolic fiber
	• 40½″ length (with end rings)
	• Aluminum end-rings and mirror cells
Mount:	• Lightweight, cast aluminum
	• Fine adjust declination control
	• 5° to 85° latitude compatibility
	• 115V ac hysteresis synchronous clock drive system

MEADE INSTRUMENTS CORPORATION TELESCOPES

Meade Instruments Corporation offers telescopes for beginners, serious amateurs, educational institutions, and research organizations. Complete information can be obtained by writing to Meade Instruments Corporation at 721 West 16th Street, Costa Mesa, CA 92627.

Meade 6-Inch and 8-Inch Reflecting Telescopes

These telescopes give large-aperture performance at a very reasonable cost. These totally integrated systems permit detailed study of the solar system and deep space, while simultaneously allowing you to become involved with long-exposure color astrophotography.

The primary and secondary mirrors of Meade's reflecting telescope models are manufactured of low-expansion, fine-annealed Pyrex brand heat-resistant glass. Each mirror is ground, polished, and figured to surface accuracies within the professional standard 1/10 wavelength of mercury green light.

The Barlow lens is convenient for increasing power when used with an eyepiece. When inserted into the telescope focuser, the Barlow doubles the power that a given eyepiece yields when used alone. A 9-mm eyepiece would give 270 × instead of 135 × in conjunction with a Barlow. A quality Barlow lens will not diminish image clarity or resolution. There are several high-performance Barlows available from manufacturers other than Meade.

Resolving power is very important to any telescope's light-gathering capabilities. Meade's 6-inch telescopes resolve star points as close as 0.74 arc seconds apart—the theoretical limit for a 6-inch telescope. Their 8-inch reflectors resolve to 0.56 arc seconds, which is again the theoretical limit.

The equatorial mounts included with Meade's 6-inch and 8-inch reflectors are made of machined aluminum and provide a lightweight, solid platform for the optical tube systems. Polar and declination shafts are of ground, turned, and polished 1-inch diameter solid steel. The polar axis of each mount incorporates a precision-ground ball bearing at the critical thrust point, easily permitting the addition of heavy cameras, guide telescopes, and other auxiliary systems. These telescopes are easy to handle even with the additional weight carried by the mounting. The machining tolerance of all mount components is .001 inch, eliminating both backlash and internal vibrations.

The Meade motor drive system is included as standard equipment on 6-inch models 628 and 645 and on 8-inch models 826 and 856. It may be purchased as an optional item on other models either at the time of the original purchase or possibly at a later date. This motor drive system will operate from any standard 115-volt ac outlet (220-volt/ 50-Hz drive systems are available for foreign operation) and will accurately track celestial objects via a synchronous timing motor actuated through a precision reduction gear system. You need not touch the telescope to follow the stars, planets, moon, or deep space objective. The drive system enables the instrument to track these objects automatically, compensating fully for the effects of the earth's rotation. An automatic clutch mechanism is included so that even when the motor drive is in operation, the telescope tube may be moved manually at will in any direction. When you release manual contact with the telescope, the drive will automatically resume tracking.

Another feature is the auxiliary right ascension control. This optional control knob is available with the motor drive at an additional cost and will permit manual centering of the telescopic image. The control is located just above the dust cover of the drive system and is easily installed (Fig. 3-14).

Two finely indexed celestial circles, one each in right ascension and declination, are provided on Meade's 6-inch and 8-inch telescopes. These setting circles facilitate the location of faint objects

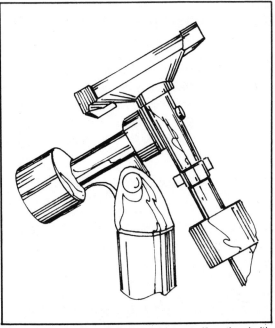

Fig. 3-14. The auxiliary right ascension control is optional with the Meade 6-inch and 8-inch reflecting telescopes.

from cataloged celestial coordinates. The standard 6-power viewfinder includes a coated, achromatic lens of 30-mm aperture, with cross hair eyepiece. The viewfinder can help you quickly and accurately locate celestial objects for detailed study in the main telescope.

The special precision rack and pinion focuser on these telescopes permits smooth, rapid focusing of the image without backlash. The tension of the focuser is fully adjustable and easily operated. The tubes used on these reflecting telescopes are of a strong, fibrous material with fine insulating characteristics. Thermal tube currents are reduced to the minimum. Each tube is finished in a white, ultrahard epoxy coating for maximum durability and is supplied with polished aluminum trim rings at each end. Cast aluminum, machined mirror cells are precisely fitted to the telescope tube. The four-vane secondary mirror supports or "spiders" are of spring steel. Once set, the secondary mirror will retain its optical collimation indefinitely with respect to the primary. It will not sag or lose centration.

Meade's 6-inch and 8-inch telescopes allow many photographic applications with both color and black-and-white film. Excellent photographs of the moon and planets will be possible on those models that are not equipped with motor drive. For longer exposures of deep space objects, you must use the motor drive. Amateurs who wish to take astrophotographs of long duration will find that Meade offers drive correctors to aid in this precise guiding. These models include prepositioned focal point settings for both first (prime) focus and eyepiece-projection photography, and adapters for virtually all 35-mm cameras are available from Meade.

All adjustments to the mounting and optics are done at the factory before shipment. Simple fine-tuning of the optics during the telescope's first use may be necessary. This type of tuning should take a few minutes and present no problems for even the first-time astronomer.

Meade 6-Inch f/8. The Meade 6-inch f/8 models combine large-aperture optics with a general-purpose focal ratio (Fig. 3-15). The 6-inch f/8 will provide more than ample capability for the majority of amateur applications.

There are three telescopes in this series. All have identical specifications and will operate in much the same manner. One is the basic telescope, another comes with Meade's motor drive, and the third is equipped with the same motor drive designed for foreign operation. All three come complete with equatorial mount, pier, and tripod legs; setting circles on both axes; number 67 rack and pinion focuser with 1¼-inch eyepiece holder; and a 6×30-mm achromatic viewfinder with cross hairs. The motor drive system includes a timing motor, reduction gear system, automatic clutch, and dust cover. The optical system consists of a 6-inch f/8

parabolic primary mirror and matching elliptical flat secondary mirror. The eyepieces are multicoated and threaded for photovisual color filters.

Meade 6-Inch f/5 Wide-Field. The 6-inch f/5 telescope has been designed specifically for the deep space observer and astrophotographer (Fig. 3-16). This telescope offers wide photovisual fields with high quality field corrections. The fast f/5 photographic speed enables deep space photography in a minimum of exposure time—a characteristic that is of particular value when using color film. This telescope will yield the wide fields of view and high levels of image brightness desirable

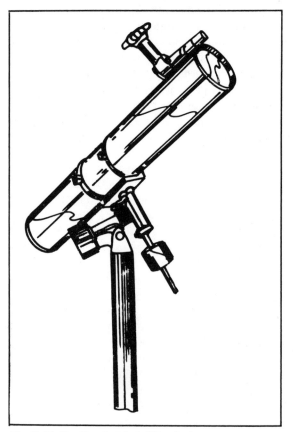

Fig. 3-15. The Meade 6-inch f/8 reflector.

Meade 8-Inch f/6. The large aperture of Meade's 8-inch telescope system enables the serious amateur to study the solar system and deep space in great detail (Fig. 3-17). Quantitatively, in the 8-inch models, almost 79 percent more light enters the telescope than in the 6-inch models, resulting in greater resolving power, brighter images, and more effective use of higher magnifications. The f/6 focal ratio allows the use of fine-grained color or black-and-white films, as well as high-quality astrophotographs of the Orion nebula, the Andromeda galaxy, and other deep sky phenomena. Despite the size of its large aperture, the Meade 8-inch reflector is a simple, trouble-free instrument to operate. Due to its 50-inch tube length, it can be easily transported to the field for viewing and study purposes.

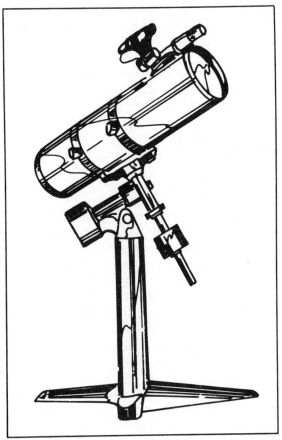

Fig. 3-16. The Meade 6-inch f/5 reflector is designed for deep space observing and astrophotography.

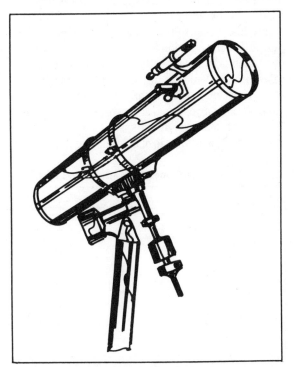

Fig. 3-17. The Meade 8-inch f/5 reflector.

in the observation of nebulae, galaxies, and other deep space phenomena.

The f/5 wide-field telescope is offered by itself or with the motor drive of your choice, depending upon American or European operation. It comes complete with all mounting components and much the same systems as Meade's other 6-inch telescopes. The only major difference is with the optical system, which is a 6-inch f/5 parabolic primary mirror with a matching elliptical flat secondary mirror.

A dual-axis corrector, photo-guide telescope, and a convenient accessory shelf are available. With any standard 35-mm camera attached to the focuser of the main telescope, you can track through the guide telescope the progress of a photographic exposure and make any necessary fine corrections with the universal joystick of the control unit. During visual observations through the main 8-inch telescope, dual-axis control permits precise electronic adjustments of the telescope's position for easy image centration in the telescopic field. The scope of this instrument's capability is only limited by the motivation of its owner.

Meade Research Series

The Meade Research Series 8-inch, 10-inch, and 12½-inch reflecting telescopes are designed for advanced amateur applications or for the college or school observatory. These precision, large-aperture instruments utilize heavy-duty components, permitting the addition of a wide range of auxiliary equipment. Included with each Research Series telescope is a listing of important standard features for serious observational and photographic work, along with many optional accessories and systems.

The telescopes in this series—in all apertures—carry a focal ratio of f/6. This focal ratio was chosen by Meade for optimum performance both visually and photographically, while permitting the large-aperture optics to be housed in relatively short tube lengths. The f/6 focal ratio permits wide-field observations of star clusters, nebulae, and other deep space phenomena. Used in conjunction with a 2 × Barlow lens for an effective focal ratio of f/12, the observation of fine planetary and lunar detail will be well within the range of these telescopes. For the astrophotographer, the f/6 focal ratio yields a fast photographic system without sacrificing field flatness and without the addition of any special focal reduction lenses.

Magnification ranges for the telescopes in this series are shown in Table 3-13. The magnification appropriate to a specific observation depends both on the nature of the object observed and on prevailing atmospheric conditions. Complete information on the proper magnifications for various astronomical objects comes with each telescope.

The two Series 2 orthoscopic eyepieces (1¼-inch outside diameter) supplied with each Research Series telescope yield wide, flat fields of view to the edge of the telescopic field. Each eyepiece includes multicoated high-transmission optics for maximum image brightness and contrast. Eyepieces are available optionally to cover virtu-

Table 3-13. Magnification Ranges for Meade's Research Series Telescopes (courtesy Meade Instruments Corporation).

Meade Research Series Telescopes: Magnifying and Resolving Powers			
Magnifying Power* when used with an eyepiece focal length of:	8" f/6 Model 880	10" f/6 Model 1060	12½" f/6 Model 1266
4mm	305X (610X)	381X (762X)	476X (952X)
6mm	203X (406X)	254X (508X)	318X (636X)
7mm	174X (348X)	218X (436X)	272X (544X)
9mm	135X (270X)	169X (338X)	212X (424X)
10.5mm	116X (232X)	145X (290X)	181X (362X)
12.5mm	98X (196X)	122X (244X)	152X (304X)
15.5mm	79X (158X)	98X (196X)	123X (246X)
18mm	68X (136X)	85X (170X)	106X (212X)
20mm	61X (122X)	76X (152X)	95X (190X)
25mm	49X (98X)	61X (122X)	76X (152X)
32mm	38X (76X)	48X (96X)	60X (120X)
40mm	30X (60X)	38X (76X)	48X (96X)
Resolving Power (arc secs.):	0.56	0.44	0.36

*Numbers in parenthesis indicate Magnifying Powers when specified eyepiece is used in conjunction with a 2X Barlow lens.

Fig. 3-18. Meade's Research Series reflectors. From left to right, the 10-inch, the 12½-inch, and the 8-inch (courtesy Meade Instruments Corporation).

ally any observing or photographic application.

Most mechanical specifications pertaining to Meade's 6-inch and 8-inch telescopes will apply to those in the Research Series. The equatorial mounting has been enlarged and strengthened to provide the additional support needed for the telescopes in this series. Oversize gear, clutch, and bearing components are used, permitting the addition of heavier auxiliary equipment without risk of strain on the mechanism. Polar and declination shafts are 1½-inch diameter, and the polar axis shaft is rigidly mounted in two preloaded ball bearings that will never require lubrication.

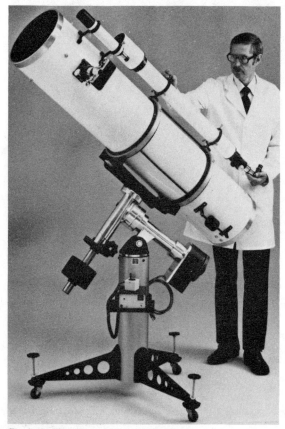

Fig. 3-19. The Meade 10-inch f/6 reflecting telescope (courtesy Meade Instruments Corporation).

As with the other telescopes offered by Meade, those in the Research Series can be had with motor drive for either American or European operation. The standard telescope in each configuration includes the motor drive.

Meade 8-Inch f/6. Figure 3-18 shows the 8-inch f/6 reflecting telescope (far right) complete with optional accessories, including a photo-guide telescope, drive corrector and accessory shelf. The 10-inch and 12½-inch telescopes are also shown. The telescope comes complete with equatorial mount, pier, and tripod legs and incorporates a fully machined tube rotation system with Teflon bearings. The optical system is an 8-inch f/6 parabolic primary mirror with a matching elliptical flat secondary mirror.

Meade 10-Inch f/6. Meade's 10-inch f/6 reflecting telescope comes complete with motor drive for either American or European operation (Fig. 3-19). This is basically the same telescope as the model just discussed, except that it is larger.

Meade 12½-Inch f/6. Meade's top-of-the-line model in the Research Series is shown in Fig. 3-18 (center) complete with optional accessories and auxiliary systems. This high-performance instrument will provide you with many years of pleasure and learning experience.

Meade Refracting Telescopes

The refracting telescopes in this series provide excellent range, clarity, and quality of viewing. The achromatic objective lenses, on these refractors have been computer-designed, resulting in clear, sharp images that are fully corrected for both spherical and chromatic aberrations. Each telescope is available with optional accessories and auxiliary systems.

Meade 2.4-Inch Equatorial Refractor. The 2.4-inch refractor shown in Fig. 3-20 will enable you to see cloud belts on Jupiter's surface, Jupiter's satellites, Saturn's ring system and satellites, lunar craters, the changing phases of Mercury

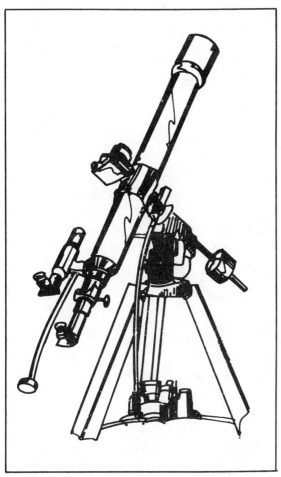

Fig. 3-20. The Meade 2.4-inch equatorial refractor.

achromatic objective resolves to the theoretical limit (1.9 arc seconds) of its aperture with excellent definition to the very edge of the field. Optional adapters for standard 35-mm cameras are available.

The 2.4-inch refractor includes equatorial mount with flexible steel cable, slow-motion controls in both declination and right ascension, locks on both axes, direct-reading latitude control, azimuth control and lock, setting circles on both axes, deluxe (coarse and fine) rack and pinion focusing mechanism with .965-inch eyepiece holder and drawtube lock, 6 ×-30-mm right-angle achromatic viewfinder with cross hairs, an air-spaced achromatic objective lens, and clear aperture 60 mm (2.4 inches). The focal length is 900 mm (f/15). A hardwood tripod and accessory tray are standard equipment with this refractor.

Meade 3.1-Inch Equatorial Refractor. You can study many celestial phenomena with this 3.1-inch refractor (Fig. 3-21). The additional light-gathering power permits observation of Jupiter's surface markings, including the Red Spot and variable characteristics of the cloud belts, the Cassini division in Saturn's rings, and prominent features of the Martian landscape. Deep sky objects take an added brilliance with the larger aperture. Galaxies and nebulae are seen in wider extension, and closer and fainter double stars are resolved.

With 178 percent of the 2.4-inch refractor's light-gathering power, this model is designed for more advanced applications. The 80-mm (3.1-inch) objective is mounted in a special collimation cell to accurately set and fix optical alignment. The resolving power of this lens is 1.5 arc seconds—the theoretical limit. Eyepieces included as standard equipment cover virtually the entire range of practical magnifications—from 54 × to 300 ×. Accessories are included.

A variation on this refractor is the model 305. This model is identical to the 3.1-inch one, except it is equipped with high-performance orthoscopic eyepieces in the standard 1¼-inch American-size

and Venus, etc. These objects are seen not as vague, fuzzy abstractions but as clear, distinct images. You can observe Uranus and Neptune, galaxies, nebulae, variable and multiple stars, and star clusters with this telescope.

The 2.4-inch telescope is a complete astronomical instrument. The three standard eyepieces cover a power range from 41 × to 150 ×. In conjunction with the 2 × Barlow lens, each of the eyepiece powers is doubled. The air-spaced barrel diameter. These eyepieces reduce eye fatigue and significantly increase the field of maximum resolution, particularly at higher powers.

The specifications for both 3.1-inch refractors are much the same as those on the 2.4-inch models, with a few minor differences. Accessories and auxiliary controls are also offered.

Meade 4-Inch Equatorial Refractor. The

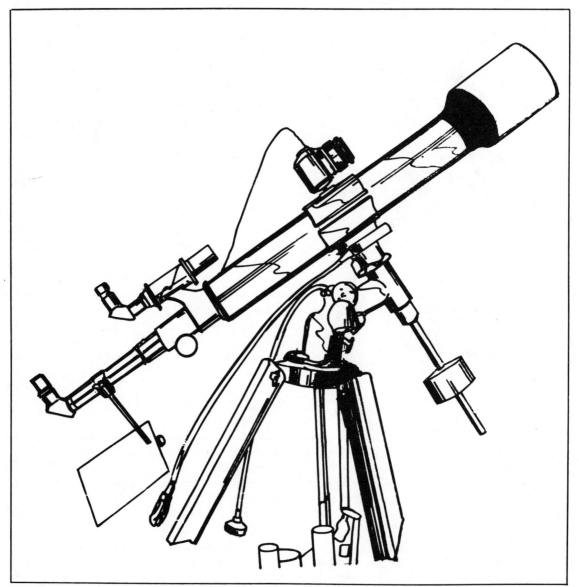

Fig. 3-21. The Meade 2.1-inch equatorial refractor.

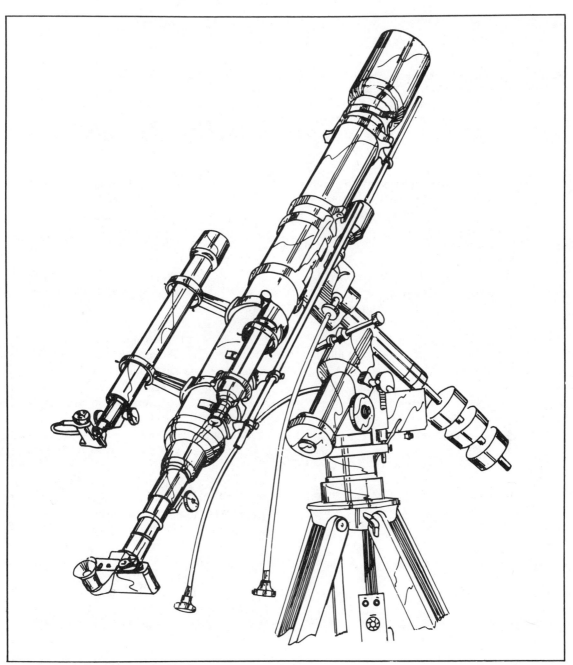

Fig. 3-22. The Meade 4-inch equatorial refractor.

4-inch refractor shown in Fig. 3-22 is excellent for the advanced amateur or for the school or college observatory. The greatly increased resolving power of the 4-inch instruments, coupled with the definition of a large, high-quality objective lens, enables you to observe thousands of double stars and deep space objects. Within the solar system, the planets present a wealth of ever-changing detail and features. The craters and seas of the moon can be seen in magnificent detail. Shadows cast by Jupiter's satellites as they transit the planet's disk are readily observed, and minor divisions in Saturn's rings are easily resolved. During favorable oppositions, significant studies of Mar's surface and atmospheric phenomena may be carried out with this refractor. The variety of color filters available will permit further planetary image enhancement.

The equatorial mounting on the 4-inch models is manufactured to accept weights and torques far greater than those required in normal applications of the telescope, thus permitting the attachment of other systems for increased usages. Full 360-degree worn gear drive systems that are enclosed and protected from dust and damage are incorporated on both axes, so that manual operation of this telescope is smooth and effortless. Electric motor drive, variable frequency correctors, color filters, astrocamera, and a guide telescope are available optional features. American-size, 1¼-inch outside diameter eyepieces combine with the 2 × telenegative amplifier to yield magnifying powers from 38 × to over 400 ×. The achromatic, air-spaced f/15 objective lenses readily resolve to the diffraction limit.

The telescopes in this series are well-suited for general terrestrial observations. Because astronomical telescopes give an inverted image, an erecting device is required to reinvert this image for nonastronomical applications. This device (called an erecting prism system) is available at an additional cost.

CROSS OPTICS TELESCOPES

Cross Optics was established in 1957 to represent a French company's products in the United States. The company, Ets. S.R. Clave, produces precision instruments. Cross Optics designs and manufactures optical components and telescopes to fulfill astronomical and industrial research requirements. For complete information on their products, write to Cross Optics at P.O. Box 27292, Escondido, CA 92027.

Nustar 14

This Newtonian reflecting telescope will permit wide-field, low-magnification studies of the sky down to richest field powers (50 × and a full degree of field). When turned to the moon, craterlets a fraction of a mile in diameter are revealed by color and albedo differences on the lunar surface that are readily visible. Planetary surfaces reveal a myriad of colors and fine structure. The four Galilean satellites are distinctly resolved and can be identified on the basis of relative size and color. Markings on Ganymede can be identified under the best conditions, as can vague albedo differences on Saturn's Titan. The three-dimensional effect of Jupiter's satellite/shadow transits, as well as Saturn's multicolored ringed beauty, will be readily visible.

When turned to the stars, the Nustar 14 will penetrate to visual magnitude 15.5 under good conditions, revealing 50 million stars in the sky. Under exceptional conditions at tall mountain sites, approximately one additional magnitude of penetration can be obtained, revealing 105 million stars in the sky.

Deep sky objects are easily found with the large field of view obtainable using the Nustar 14. The largest open star clusters can be seen almost entirely in one 50 × (1-degree) field. The globular clusters M13 and M3 appear more impressive than their photographs at 350 ×. The spiral arms of galaxies such as M33, M51, and M101 will be visible, along with the larger star clouds in M31 and M33. Even the brighter quasars can be seen.

The primary mirror used in the Nustar 14 is fine-annealed 14.5-inch diameter lightweight Pyrex brand heat-resistant glass. Its surface is aluminized and overcoated with silicon monoxide for long life. The focal ratio is f/4.17, and the focal length is 60 inches. The secondary diagonal mirror is also fine-annealed Pyrex brand heat-resistant glass with a minor axis diameter of 3.5 inches. The mirror is mounted in an adjustable stress-free cell precisely located by a four-vane spider support.

American standard 2-inch and 1.25-inch oculars are accepted by the Nustar 14 telescope. High-quality eyepieces should be used.

The rotating tube on the Nustar 14 can be rotated + 45 degrees from its mean position. A convenient eyepiece position can thus be obtained. The rotation mechanism is free of play, and the tube is attached rigidly to the saddle. The telescope tube has a porcelain off-white exterior and flat black interior. The tubing used is rigid, thermally non-conducting, round, and lightweight.

The Nustar 14 has a torque tube-type equatorial mount of the kind used on major observatory telescopes. This mounting is very strong and rigid. It permits tracking on objects past the observer's meridian. The main components are precision-machined from a stable aluminum alloy. Declination and polar shafts are machined precisely 90 degrees to each other, which is essential for accurate locating and tracking. The trapezoidal base has four large casters. Two casters are designed to swivel. This permits considerable flexibility during setup and storage. Four leveling screws are attached to the trapezoidal base. This equatorial mount is full-adjustable for latitude.

The motor drive system on the Nustar 14 is called a chain drive and provides economy, simplicity, and extremely high accuracy. This drive is widely used in many large observatories.

A high torque synchronous motor turns through precision reduction gears to a shaft on which two small sprocket gears ride. Two parallel rollerless chains ride on these gears, which go around a large, toothless drive disk attached by a clutch to the polar axle. This type of drive can be "tuned" to track with a periodic error smaller than 1 arc second peak-to-valley. This type of accuracy will permit convenient tracking for astrophotography.

The setting circles on the Nustar 14 are a full 10 inches in diameter, making them easy to read. Each has a black background with white-filled calibration divisions and numbers. The hour angle circle functions as a slip ring on the drive gear (disk), so one setting on a known star at the beginning of the evening permits direct reading thereafter.

A manual slow motion control for declination movement uses a large tangent arm and precision screw mechanism. This produces a smooth, responsive motion. An optical frequency generator is available for electronic slow motion control of the right ascension motion by hand paddle control. Electric declination slow motion control is a sepa-

Table 3-14. Specifications and Other Information for the Nustar 14 Newtonian Reflector (courtesy Cross Optics).

- Clear aperture: 14.4-inches (366mm)
- Optical plan: Newtonian
- Exceeds Rayleigh criterion and Dawes' limit
- High-contrast images
- 1.0-degree field at 50 ×
- High resolution at 760 ×
- Limiting magnitude: Fainter than 15.5 visually
- Optical tube length: 65 inches
- Photographic speed: Fast, f/4.2
- Strong torque-tube type equatorial mt.
- Heaviest component: 52 pounds
- Extremely accurate sidereal drive (10 × smaller periodic error than worm gear)
- 10-inch diameter setting circles
- Designed and built by Cross Optics
- 6 to 8 weeks delivery
- 35° ± 15° N/S latitude adjustment, with extendable range and reversable synchronous motor provide virtually worldwide operation
- Special-order compact f/12 Cassegrain and f/9 Ritchey-Chrétien versions of Nustar 14 available on a limited basis
- Most performance/cost-effective telescope
- World's largest aperture transportable production telescope

rate option. Both declination and right ascension electric slow motions are available with control by a single hand paddle.

Although this telescope is quite large, it is easily transported and can be disassembled by one person in less than 5 minutes. Complete specifications and other information are listed in Table 3-14.

Custom Nustar 14

Custom Nustar 14 telescopes offer much the same degree of optical and mechanical stability produced on the standard Newtonian Nustar 14. The Cassegrainian optical plan of this model permits greater compactness for ease of transport. The optical path is folded by two additional elliptical flat mirrors to bring the Nasmyth/Coude focus to a fixed eyepiece position at the north end of the polar axis. From this position, you view through the telescope eyepiece or accessory as in an inclined microscope. These specially designed telescopes are available in a limited number. Table 3-15 provides complete specifications on this custom telescope.

Table 3-15. Specifications for the Classical Cassegrain Nustar 14 and the Ritchey-Chretien Nustar 14 (courtesy Cross Optics).

Specifications	Nustar 14 Classical Cassegrain	Nustar 14 Ritchey-Chretien
Primary Mirror Diameter	14.5"	14.5"
Clear Aperture	14.4"	14.4"
Light Grasp (Compared to 8mm pupil)	1911 ×	1638 ×
Focal Length	216"	130"
Useful Magnification	73 ×-1100 ×	73 ×-1100 ×
Resolution: Visual	.32 arc sec	.32 arc sec
Resolution: Photographic	120 lines/mm	200 lines/mm
Stellar Magnitude Limit, Visual	15.6	15.4
Number of Stars Visible (approximately)	52 million	44 million
Photographic Speed	f/15	f/9
Image Scale (Field of View)	.27°/inch	.44°/inch
Field of View (Visual 30mm FL Plössl)	.29°/inch	.48°/inch
Unvignetted Field	1.5°	1.5°
Secondary Obscuration	4.2"	6.7"
Finderscope	8 ×-50 mm	8 ×-50 mm
Suggested Eyepieces	75 mm-73 ×	45 mm-73 ×
	40 mm-137 ×	2X Barlow
	30 mm-183 ×	25 mm-132, 264 ×
	20 mm-274 ×	16 mm-206, 413 ×
	16 mm-342 ×	10 mm-330, 660 ×
	12 mm-456 ×	6 mm-550, 1100 ×
	8 mm-686 ×	
Setting Circles Diameter RA	10"	10"
DEC	10"	10"
Drive Disk/Gear Diameter	10.6"	10.6"
Clock Power (110v, 60Hz)	5 Watts	5 Watts
Mount, Equatorial, Type	Torque Tube	Torque Tube
Optical Tube Length	42"	38"
Weight	175 lbs	175 lbs
Heaviest Component	48 lbs	50 lbs
Instrument Payload Capacity (at Nasmyth/Coude focus)	50 lbs	50 lbs

Table 3-16. Specifications for the Classical Cassegrain Nustar 24 and the Ritchey-Chretien Nustar 24 (courtesy Cross Optics).

Specifications	Optical Plan	
	Classical Cassegrain	Ritchey-Chretien
Primary Mirror Diameter	24.2″	24.2″
Clear Aperture	24.0″	24.0″
Light Grasp (Compared to 8 mm Pupil)	5,193 ×	4,587 ×
Focal Length	122-1500 ×	216″
Useful Magnification	122-1500 ×	122-1500 ×
Resolution: Visual	0.19 arc sec	0.19 arc sec
Resolution: Photographic	120 lines/mm	200 lines/mm
Stellar Magnitude Limit, Visual	16.7	16.6
Number of Stars Visible (approximately)	105 million	99 million
Photographic Speed	f/15	f/9
Image Scale (Field of View) ••	.16°/inch	.27°/inch
Field of View (Visual 30 mm FL Plossl)	10.5 arc min	17.5 arc min
	(.18°)	(.29°)
Unvignetted Field	1.5″ (.24°)	4″ (1.06°)
Secondary Obstruction	7.8″	11″
Finderscope	---------8 ×-50 mm, 23 ×-152 mm---------	
Suggested Eyepieces	75 mm-122 ×	45 mm-122 ×
	40 mm-229 ×	25 mm-219 ×
	30 mm-305 ×	16 mm-343 ×
	20 mm-457 ×	12 mm-457 ×
	16 mm-572 ×	10 mm-549 ×
	12 mm-762 ×	8 mm-686 ×
	10 mm-914 ×	6 mm-914 ×
	8 mm-1143 ×	5 mm-1047 ×
Setting Circle Diameter RA	10″	10″
DEC	10″	10″
Drive Disk/Gear Diameter	21.2″	21.2″
Clock Power (110v, 60Hz)	10 Watts	10 Watts
Mount, Equatorial, Type	Torque Tube	Torque Tube
Optical Tube Length	80″	70″
Weight	700 lbs	700 lbs
Heaviest Component	175 lbs	175 lbs
Instrument Payload Capacity	100 lbs	100 lbs

Nustar 24

The Nustar 24 telescope is a moderately large aperture telescope for research and instruction. By utilizing modern lightweight optical technology and manual slewing, substantial cost savings are realized while maintaining research level performance. Nustar 24 telescopes are well-suited for the private observatory or university. Trailer mounting for the telescope is an available option. Specifications and other pertinent information are given in Table 3-16.

OPTICAL TECHNIQUES, INC. TELESCOPES

Optical Techniques, Inc. was formed by two former executives of Questar Corporation—John Schneck and Robert Richardson. They have produced a Maksutov-Cassegrain telescope at a reasonable price. Schneck and Richardson have developed a modern catadioptric optical system with excellent performance characteristics.

The optical system utilized in the Quantum Series of Maksutov-Cassegrains provides maximum contrast and resolution in a commercially produced catadioptric telescope. Instead of the usual f/2 primary mirror used in compound telescopes, the Quantum's primary is of f/2.5 ratio. This results in several advantages.

The diffraction-limited field is much larger—a full 9 mm in diameter for the Quantum Four and 12 mm for the Quantum Six. Secondly, a smaller secondary mirror spot can be employed on the inside of the corrector plate. Their central obstructions are only 33 percent of their clear apertures.

The Quantum Four and Six, with their smaller central obstructions, image nearly 68 percent of the light into the central Airy disk. Only 32 percent of the light goes to the contrast-destroying diffraction rings. This significant improvement in contrast, coupled with the diffraction-limited resolution, results in a catadioptric telescope with the capability to reveal delicate, low-contrast detail on the moon, planets, and deep sky objects.

In addition to giving the telescopes a sculptured appearance, the mounting system allows for much greater flexibility in the design, selection, and use of accessories. It also permits more comfortable north-sky viewing and a freedom of operation. Although these instruments are quite portable and lightweight (the Quantum Four weighs 14 pounds, and the Quantum Six weighs 30 pounds), they are also quite stable. Thrust bearings have been incorporated on both axes. The optical tube assembly dismounts instantly with a single knurled knob, so it can be used separately as a terrestrial telephoto lens. An astrographic camera platform can take the place of the optical barrel for wide-angle, deep sky photography.

Other features of this mounting include an electric gear-driven sidereal clock drive located in the mounting base; setting circles that can be comfortably read from a viewing position; and smooth, continuous 360-degree manual slow-motion controls in both axes. Legs are also available so that the instruments can be operated in equatorial position

from any suitable tabletop or flat surface. The telescope can be mounted to any stable tripod through the ¼-20 threaded socket located on the mounting's base.

There are two control levers located at the rear of these instruments. One activates a built-in 1.75 × Barlow lens, while the other selects the option of right-angle viewing for normal observing or straight-through viewing for photography. For photographic work, the central cover cap at the back of the telescope is removed. A camera swivel coupling is attached in its place. A camera with the appropriate T-ring is then attached to the coupling. The eyepiece will remain at its normal position even with the camera attached.

The focusing system on the Quantum Series of telescopes is a bit different than that of most other instruments. Movement of the primary takes place by turning a focus knob located on the right side of the barrel toward the rear. This side position makes operation positive and natural without introducing any difficulties when adding accessories. One notable design feature is a total absence of image shift during focusing, which results in great accuracy and reliable performance.

The Quantum Four accepts standard 1¼-inch outside diameter (O.D.) eyepieces, while the Quantum Six accepts 2-inch outside diameter oculars and is furnished with a 1¼-inch adapter. Optical Techniques, Inc. supplies the University Optics 16-mm multicoated Konig eyepiece, which has crystalline imaging and an 80-degree apparent field. With the built-in Barlow, the 16-mm ocular field yields powers of 95 × and 166 × on the Quantum Four and 140 × and 250 × on the Quantum Six. The company utilizes a precision flat-surface mirror for the built-in star diagonal. It yields better image quality than that provided by a prism.

Other standard features of the Quantum instruments include high transmission AR coatings on the corrector lens and enhanced aluminum coatings on the primary, secondary, and diagonal mirrors. Each instrument is also equipped with a right-angle finder of large aperture—a 6×30 on the Quantum Four and a 8×50 on the Quantum Six.

Quantum Four, Six, and Eight Telescopes

The Quantum Four is a complete 4-inch aperture, f/15, Maksutov Cassegrain telescope (Fig. 3-23). This instrument weighs only 15 pounds.

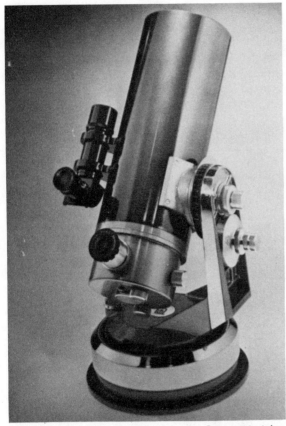

Fig. 3-23. The Quantum Four Maksutov Cassegrain telescope (courtesy Optical Techniques, Inc.).

Standard equipment includes an optical tube assembly with built-in star diagonal and 1.75 × Barlow lens, 16-mm F.L. (focal length) Brandon eyepiece (95 ×-166 ×) of 50 degrees of apparent field, 6×30-mm right angle finder, enhanced silver-coated optics, single arm fork mount with setting circles, slow motion controls, electric clock drive, lens cap, carrying case, and instruction booklet.

The Quantum Six is also light and portable, weighing in at approximately 35 pounds (Fig. 3-24). Standard equipment is basically the same as that of the Quantum Four, except that the eyepiece is 140 ×−250 × of 50 degrees of apparent field—8×50 mm.

Optical Techniques, Inc. offers many optional accessories. Special adapters make it possible to mount any 35-mm camera easily.

The Quantum Eight shown in Fig. 3-25 is an 8-inch telescope utilizing a Maksutov Cassegrain optical design. It is sold as an optical tube assembly

Fig. 3-24. The Quantum Six Maksutov Cassegrain telescope (courtesy Optical Techniques, Inc.).

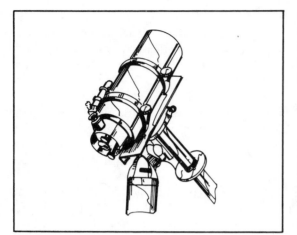

Fig. 3-25. The Quantum Eight Maksutov Cassegrain telescope.

only and is intended for incorporation into a separate mounting and observatory configuration. The equatorial mount, oculars, and other accessories are available from Optical Techniques, Inc. Specifications are given in Table 3-17.

Quantum 100 Series

The 100 Series models, with the number denoting the aperture in millimeters, are the Quantum 100 and the Quantum 150. Each consists of an unmounted barrel assembly without control box, built-in Barlow lens, or diagonal. The observation function proceeds via optional, detachable appliances that fit into the 1¼-inch accessory adapter. This adapter accepts an eyepiece alone, a porro prism and ocular, a plane diagonal and eyepiece, or a

Table 3-17. Specifications for the Quantum Four, Six, and Eight (courtesy Optical Techniques, Inc.).

Item	Quantum Four	Quantum Six	Quantum Eight
Clear Aperture	4″	6″	8″
Focal Length	60″, f/15	90″, f/15	120″, f/15
Resolving Power	1.14 arc sec.	.76 arc sec.	.57 arc sec.
Diffraction-Limited Field	9 mm. dia.	12 mm. dia.	15 mm. dia.
Central Obstruction	33 mm. (33% diam. 10% area)	50 mm. (33% diam., 10% area)	67 mm. (33% diam., 10% area)
Baffle System	Double (Primary & Secondary)	Double (Primary & Secondary)	Double (Primary & Secondary)
Unvignetted Field	23 mm. dia.	35 mm. dia.	40 mm. dia.
Photographic Field of View	.94° per inch	.58° per inch	.47° per inch
Limiting Magnitude	13.4	14.3	14.8
Near Focus	15′	30′	60′
Primary Mirror	Pyrex, 4.45″ dia., f/2.5	Pyrex, 6.65″ dia., f/2.5	Pyrex, 8.85″ dia., f/2.5
Corrector Lens	BK-7, 4.18″ dia.	BK-7, 6.25″ dia.	BK-7, 8.35″ dia.
Coatings			
Reflective Surfaces	ALSiO	ALSiO	ALSiO
Corrective Lens	AR-MgF$_2$	AR-MgF$_2$	AR-MgF$_2$
Barlow Lens	1.75 ×	1.75 ×	1.75 ×
Eyepiece holder	1.25″ I.D.	2″ I.D. with 1.25″ adapter	2″ I.D. with 1.25″ adapter
Ocular	16 mm. F.L. 1.25″ O.D.	16 mm. F.L. 1.25″ O.D.	16 mm. F.L. 1.25″ O.D.
Power	95 × and 166 ×	140 × and 250 ×	190 × and 333 ×
Finder (Right Angle)	6×30	8×50	8×50
Mount Type	Single-Strut Fork	Single-Strut Fork	N.A.
Dec. Thrust Bearing	3.0″ dia.	4.125″ dia.	N.A.
Dec. Circle	3.0″ dia. — 2° divisions	4.125″ dia. — 2° divisions	N.A.
R. A. Thrust Bearing	3.63″ dia.	5″ dia.	N.A.
R. A. Circle	6.5″ dia., — 1° divisions	8.25″ dia. — 1° divisions	N.A.
R. A. Gear	4″ dia.	6″ dia.	N.A.
Drive Motor	Synchronous	Synchronous	N.A.
	110V, 60Hz, 2.7 watts	110V, 60Hz, 2.7 watts	N.A.
Height (Upright)	17.5″	25.5″	N.A.
Weight	15 pounds (approx.)	35 pounds (approx.)	N.A.

T-mount swivel camera coupling to hold a 35-mm SLR camera body.

Integral with each barrel is a mounting bracket with a ¼-20 tapped hole to facilitate secure attachment to a sturdy tripod. This also forms the basis for attachment of an optional motor-driven equatorial mount.

Focusing is accomplished by varying the separation between the primary and secondary mirrors. Movement of the primary takes place by turning a focus knob located at the rear of the telescope. Bothersome image shift is totally absent.

Quantum 100. Optical performance is excellent with the Quantum 100 (Fig. 3-26). As viewed through this telescope, the moon reveals such features as the Teardrop Rill near the Straight Wall. The traceries of the rill complex at Triesnecker are clearly visible. Tiny craterlets in the floor of Plate are also revealed plainly.

Jupiter presents a spectacle consisting of the four Galilean moons whose occasional shadow transits and eclipses are easily followed. Belt detail on the planet will show a changing variety of festoons, white spots, and subtle gradations of color. The famous Red Spot will be seen quite prominently when its side of the planet faces you.

Saturn's belts are much less conspicuous than Jupiter's, but the 100 will reveal them and an occasional white spot. The planet's main feature is the beautiful ring system split by Cassini's division. Attentive observers can note the difference in brightness between ring A (outer) and ring B (inside the division). On nights of good transparency, Titan and four of the fainter moons lie within the 100's reach.

More than 800 galaxies lie within reach of the 100. Emission nebula showpieces such as M-42 (Orion nebula), M-20 (Trifid nebula), and M-8 (lagoon nebula) can be seen with a rich complexity that often eludes long exposure photographs.

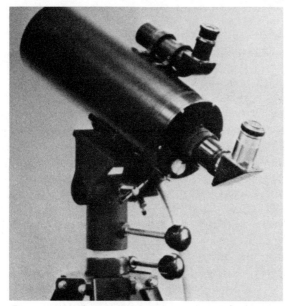

Fig. 3-26. The Quantum 100 (courtesy Optical Techniques, Inc.).

Fig. 3-27. The Quantum 150 (courtesy Optical Techniques, Inc.).

Table 3-18. Specifications for the Quantum 100 and the Quantum 150 (courtesy Optical Techniques, Inc.).

ITEM	QUANTUM 100	QUANTUM 150
Clear Aperture	4″ (100 mm.)	6″ (150 mm.)
Focal Length	60″,f/15	90″, f/15
Resolving Power	1.14 arc sec.	.76 arc sec.
Diffraction-Limited Field	9 mm. dia.	12 mm. dia.
Central Obstruction	33 mm. (33% diam., 10% area)	50 mm. (33% diam., 10% area)
Baffle System	Double (Primary & Secondary)	Double (Primary & Secondary)
Unvignetted Field	23 mm. dia.	35 mm. dia.
Photographic Field of View	1.3° (35 mm.)	1.6° (70 mm.)
Limiting Magnitude	13.4	14.3
Near Focus	12.5′	30′
Primary Mirror	Pyrex, 4.65″ dia., f/2.5	Pyrex, 6.65″ dia., f/2.5
Corrector Lens	BK-7, 4.18″ dia.	BK-7, 6.18″ dia.
Coatings		
Reflective Surfaces	Enhanced Aluminum	Enhanced Aluminum
Corrective Lens	AR-MgF$_2$	AR-MgF$_2$
Weight	5.5 pounds (approx.)	12 pounds (approx.)

Quantum 150. With the Quantum 150 shown in Fig. 3-27, you have more light and resolution. The telescope can examine more than 60 craterlets in the floor of Clavius and extremely delicate rill structure in the floor of Alphonsus. The Straight Wall begins to reveal its actual lack of straightness, and the tiny craterlets near the Teardrop Rill make their appearance plainly.

On Mars during opposition, the dark collar around the melting polar cap is clearly discernible, as are the smaller light and dark markings like Juventae Fons and Trivium Charontus. Increased richness of color variety is evident on Jupiter when looking through the 150. There is greater definition of higher latitude belt detail and the finer festoons in the equatorial region.

Saturn's ball, as seen through the 150, exhibits increased rendition of the single belt usually visible and the greater variety of hues associated with the additional light grasp. The view of the ring system includes the inner crepe ring, Cassini's division, and the evanescent Encke division in the outer A ring. Titan begins to show its disk, and Rhea, Tethys, Dione, Lapetus, and Mimas are visible under steady transparent conditions.

The 2.25-times increase in light grasp places 2000 galaxies within potential reach of the 150 under a dark, transparent sky. Fainter planetary nebulae such as M-76 (12.2 magnitude) and M-97 (12.0 magnitude) (the Owl nebula) are easily seen. The appearance of the brighter showpieces is enhanced, and their structure is better revealed by the gain in resolution. The Dumbbell nebula (M-27) appears to float amid a faint dusting of 12.0 magnitude and fainter stars.

Table 3-18 provides complete specifications and other pertinent information on both the Quantum 100 and Quantum 150 telescopes.

Chapter 4

Our Solar System

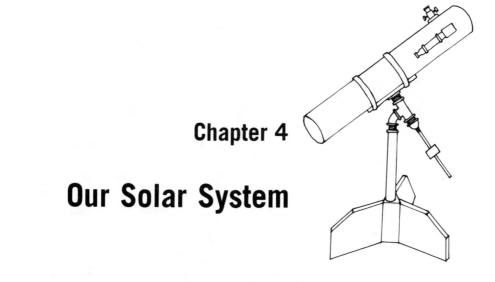

OUR SOLAR SYSTEM HAS NINE PLANETS THAT revolve around the sun. The planets that will be of most interest to the amateur astronomer are Mars, Jupiter, and Saturn. Venus and Mercury can certainly be viewed through inexpensive telescopes, but the detail offered by the three planets is usually far better. The most interesting sight for many astronomers is Saturn's rings, which can be clearly defined on a good viewing night by using a simple telescope.

The solar system consists of planets, comets, meteors, etc., that are directly affected by the gravitational field of the sun, and each travels in some form of path or orbit around it.

Some planets, including Earth, have their own satellites or moons that are in orbit around the planet. Jupiter has four moons, while Saturn and Neptune each have one. There are at least 30 smaller moons orbiting the various planets. Some 2000 asteroids are in orbit around the sun, with the largest number being located in an asteroid belt between Mars and Jupiter. Sometimes an asteroid enters an orbit around an individual planet and takes on the same characteristics of a moon.

The two planets located closer to the sun than Earth, Mercury and Venus, are commonly termed the inferior planets. Those planets with orbits outside Earth's orbit are called superior planets. Each planet's velocity is determined by its distance from the sun. Those planets farthest away from the sun travel at a much slower rate of speed. Because Mercury and Venus are orbiting within Earth's orbit, they are normally observed either in the early evening or early morning.

THE SUN

The sun is the nearest star to earth—a mere 93 million miles away. It is essentially a hot, gaseous body that produces its own energy and radiates it outward in a type of nuclear reaction, very much similar to that of a hydrogen bomb. The sun is composed of several gases. Hydrogen is found in the greatest quantity, and there are smaller amounts of helium and other elements.

If you want to observe the sun through a telescope, exercise great care. The energy the sun projects is so strong that permanent damage to the eyes may occur. Special filters are generally provided with a telescope designed specifically for viewing the sun. A properly attached filter will reduce the light intensity, but it will have no effect on the heat projected by the sun. The heat may crack the filter, eliminating its use as an eye protector. A sun diagonal will safely protect the eyes. It utilizes a flat wedge of glass, which is a flat prism arranged much like a star diagonal. The sun diagonal passes the majority of the sun's rays through the wedge, reflecting only a small amount to you.

Another safe method of observing the sun is by using a sun projection screen available through telescope manufacturers as an accessory. The telescope is turned to point at the sun by sighting along the tube through half closed lids or by looking at the sun through a double thickness of overexposed and developed photographic film. Hold a piece of white cardboard several inches from the eyepiece. Adjust the telescope until the sun's image is visible on the card. If viewing the sun in this manner, the projected image will be somewhat inferior. Because the resultant heat buildup in the scope may damage the secondary mirror and eyepiece, projection time can be as little as 60 seconds.

The safest type of solar filter is the glass window type that slips over the front cell of a telescope. These filters are designed to reduce the intensity of the solar radiation to 1/100 of 1 percent at all wavelengths. Precautions should be observed:

☐ Place the filter snugly over the telescope's front cell.

☐ Do not leave your scope unattended during an observing session.

☐ Always cap the finderscope so the heat from the sun's rays doesn't damage the delicate cross hairs.

THE MOON

The moon is the earth's natural satellite and our nearest celestial neighbor. Its landscape affords you, the viewer, with a diversified pattern of craters, cliffs, mountains, etc.

The moon is best viewed during its partial phases and at its highest point in the sky. Broad areas along the terminator will be extremely sharp and detailed at low power. It will be possible to watch the terminator gradually advance at high power. Towering peaks will blaze in sunlight, while their lower surroundings are still lost in lunar predawn shadow.

Tiny craterlets can be seen peppering the floors of large craters and walled plains like Plato and Clavius. About two days past the First Quarter, you can study the intricate terracing of the walls of the crater Copernicus. A winding chain of tiny craters will be easily visible just northwest of this crater.

Because the moon rotates on its own axis at the same rate at which it orbits the earth, one side of the moon is never visible to you. This is not one complete half of our satellite due to an effect called *libration*, which is a rocking motion of the moon. As this rocking occurs, the visible portion of the moon is changed slightly, so you can see some of the far side at different periods. This rocking motion may be caused by variations in orbital speed, or possibly by the effect of its elliptical-shaped orbit and its axis of rotation not being exactly in line with the plane of the orbit.

The reason for viewing the moon in its partial phases is there will be very little glare from reflected sunlight. You can observe the moon in other phases with special filters that reduce excess light and increase contrast.

Information on when lunar phenomena are to occur can be obtained from various astronomical almanacs. These almanacs can accurately predict such things as eclipses and occultations. An *occultation* occurs when the moon passes between the observer and a star or planet. This is a very interesting event to observe through a telescope, as it clearly shows the absence of atmosphere around the moon.

An eclipse occurs when the earth, moon and sun are directly lined up. A lunar eclipse occurs when the earth's shadow intercepts the moon's surface. Lunar eclipses are less common than solar eclipses due to the geometric angles involved, but they are visible over the whole hemisphere of the earth that is turned toward the moon.

The most conspicuous features of the moon's surface viewed through a telescope will be its thousands of craters that are of numerous sizes and shapes. You can observe the moon's "seas." These "*seas*" are not really filled with water, but they appear as rolling plains that are somewhat darker than their surroundings. The moon's surface is quite dark and will appear to be almost black.

MERCURY

Mercury, the closest planet to the sun, is a difficult planet to observe through the telescope. It revolves around the sun in 88 days. Mercury is best observed at eastern elongation in spring just after sunset or at western elongation in the fall just before sunrise. Even when Mercury can be viewed, the results may be disappointing due to the glare reflected by the sun. Mercury will merely appear as a fuzzy white disk through a telescope, and astronomers have been unable to learn very much about the surface of this fiery planet.

Mercury has no appreciable atmosphere, and it receives many more times the amount of radiation than Earth does. Because its day is equal to 88 Earth days, one side of Mercury is exposed to the sun for this period. The other side gets no heat at all. Mercury's surface is exposed to considerable temperature extremes during these periods. Although it it known that Mercury's exterior is made up of a similar rock as that of the moon, it is not known whether its inner core is solid or liquid.

VENUS

Venus is so bright that it can be clearly observed by the naked eye even in daylight hours. Venus is sometimes referred to as Earth's sister planet, because it almost matches the Earth in size and period of revolution. It also has its own atmosphere. Due to Venus's close proximity to the sun, its surface temperature is much hotter and is estimated to be about 90 times that of the Earth. The surface is maintained at this high degree of heat by an intensely thick atmosphere of clouds. The composition of the atmosphere is still unknown.

Because of these clouds, you may find that the view is disappointing, as was the case with Mercury. Venus is almost continually shrouded in dense clouds, so it will appear merely as a bright disk with no surface detail visible.

MARS

Mars is our closest neighbor among the planets. Mars is called the red planet because of its distinctive reddish hue. You can obtain detailed views of Mars's surface when it is near opposition, which is the point in its orbit when it lies at a 180-degree angle from the sun. This occurs approximately every 26 months. Mars has two small satellites that are assumed to be asteroids. These are only visible through very large telescopes.

Like other planets, Mars is best viewed when highest overhead and at moderate magnification. Mars is the only planet with permanent, recognizable surface details. You can observe enormous dust storms changing these surface details from the easily identifiable features shown on Martian maps into a variety of new forms. Orange or red filters will reduce glare and help increase the contrast of Martian surface features. A blue filter is useful for emphasizing Mars's atmospheric features.

Mars's surface is similar to the moon's in that it has many craters. Mars also has volcanoes, canyons, and other interesting landmarks that indicate volcanic action and meteor bombardment.

The polar caps and reddish color of Mars are prominent. Some areas seem to be of a greenish color, which has led some to believe there is some form of plant life or vegetation on Mars. The reddish areas are thought to be deserts. You will probably only see the polar caps of Mars with a small telescope. If detailed views are desired, a 6-inch refractor can be used to obtain more distinct observations.

JUPITER

Jupiter is the largest planet in our solar system—some 1300 times larger than Earth. It has 12 known moons or satellites, and four can be clearly seen through a telescope. Because Jupiter rotates on its axis in a much shorter period than any other planet, its shape will appear slightly more elliptical and somewhat flattened at the poles. Jupiter's belts of clouds comprise its outermost atmosphere. This atmosphere is composed mainly of hydrogen and smaller portions of ammonia, water, and methane. The temperature in this outer atmosphere is extremely cold. The methane and ammonia crystallize, resulting in changing movements and colors.

The most noticeable feature of these cloud formations, and the one that remains fairly constant, is the great Red Spot. This is an extremely large mass that appears to be brick-red in color. It spans a distance about 7,000 miles wide and 30,000 miles long. This spot may be shrinking somewhat. The spot is part of the cloud belts around the planet.

When observing Jupiter through a telescope, you can focus on one particular patch of color and follow its motion and color changes. The cloud belts rotate at a faster rate than does Jupiter itself.

Some of Jupiter's moons swing out away from the planet by as much as 10 million miles, while others come as close as 112,000 miles to Jupiter. The four largest moons are Io, Europa, Ganymede, and Callisto. Both Ganymede and Callisto are larger than our moon, and Io and Europa are somewhat smaller. These moons will sometimes appear in a straight line, because they orbit in virtually the same plane as Jupiter's equator. Sometimes a moon will not be visible because it is passing behind or in front of Jupiter.

Jupiter's larger moons are rocky and have many of the surface features as our moon. Ganymede has been observed to have huge craters and ice fields. The moon Io has active and inactive volcanoes.

A blue filter on a telescope will greatly enhance the contrasts of Jupiter's cloud belts and the Red Spot. Unwanted glare will be reduced. You can detect some detail within this Red Spot under the right conditions.

SATURN

Like Jupiter, Saturn will be visible for many months each year. Jupiter and Saturn are almost the same size, with Saturn being slightly smaller. Hydrogen dominates its atmosphere, and there are amounts of ammonia and methane. The atmosphere is not as dense, making Saturn much lighter when compared to Jupiter. The cloud belts around Saturn will be quite colorful.

Saturn's rings are clearly visible through a telescope and present gradually changing appearances at different times in Saturn's orbit around the sun. The rings may be composed of small particles, possibly frozen ammonia crystals, that are held in place by Saturn's powerful magnetic force. These rings are not solid. There are at least three definite rings. Between the A and B rings is a gap known as Cassini's Division. The third ring, which is the one closest to the planet, has been named the Dusky

Ring or the Crepe Ring. Its particles are very thinly spread out. This ring is most easily visible by the shadow it casts across Saturn's surface.

Saturn has at least 13 satellites. The largest, Titan, is almost as large as Saturn and has a similar atmosphere.

When viewing Saturn, the major divisions of the rings will be quite distinct and obvious. The surface detail of Saturn can be seen clearly under proper conditions. There is a striking amount of banded detail on the globe, which usually includes one or more white spots. There will also be a hint of belt structure near the polar region. Titan appears as a disk.

URANUS

Uranus is rarely, if ever, visible to the naked eye. Uranus is almost four times as large as Earth and has a unique orbital path. Its axis of rotation lies almost in its orbital plane inclined at an angle. This implies a retrograde motion that has long puzzled astronomers. Uranus is physically similar to both Jupiter and Saturn in that it has a thick, dense atmosphere composed of the same elements but in differing amounts. Because the surface temperature is quite cold, there is little if any ammonia present in the atmosphere. Due to the increased amount of methane present in its atmosphere, Uranus gives off a bluish-green hue when observed.

Researchers feel that Uranus has additional moons yet to be discovered. Oberon, Umbriel, Titania, Ariel, and Miranda are known moons.

Uranus has a gravitational pull of its own, but it is very weak when compared with that of Earth. It takes Uranus approximately 84 Earth years to orbit the sun. Uranus may be observed through a telescope without much difficulty, but its moons will generally not be visible except through the larger, more expensive telescopes.

Astronomers were surprised to discover that Uranus, like Saturn and Jupiter, also has rings. The rings were discovered when the astronomers were observing an occultation of a star at a time when it was known that Uranus would eclipse it. Uranus has at least nine rings at this writing. Because the rings are as brilliant and shiny as Saturn's and take on somewhat unstable shapes and motions, some people believe they may be gaseous rather than of frozen particles. Others feel they may be pieces of rock. Still others believe there are undiscovered satellites located on the sides of the rings causing erratic movements and unusual and varied shapes. Because of its great distance from Earth, very little is known about the surface of Uranus. Some scientists contend that Uranus, like Jupiter and Saturn, may not have a solid surface.

NEPTUNE

When astronomers first observed Uranus and were puzzled by their inability to track it accurately, they deduced that its path was being affected by another planet located beyond it. Neptune is not visible to the naked eye and can only be seen with a quality telescope. Neptune will appear to be smaller than Uranus and has the same bluish-green tint, but it is almost the same size. Neptune appears smaller because it is so much farther away.

Neptune has at least two moons—Triton and Nereid. Triton, the largest, circles Neptune backwards or opposite the direction in which Neptune spins. Some believe that this satellite may have collided at some time with another, resulting in the moon being thrown out of its orbit.

Neptune's atmosphere is composed of a mixture of gases. The most prominent is methane. Very little is known about Neptune's surface.

PLUTO

Many astronomers believe that Pluto was once a satellite of Neptune along with its moon, Charon, which escaped from the orbit of Neptune and settled into an orbit around the sun. Others believe they have always existed in the same manner as now.

Pluto is closer to the sun than Neptune at certain periods. This is caused by Pluto's elliptical orbit. It takes Pluto approximately 248 Earth years to revolve around the sun.

Charon was not discovered until 1978. When first noted, Charon appeared almost blurred with Pluto, making it appear as if Pluto was changing shape.

As with Neptune and Uranus, there is very little solid information available on Pluto's surface and atmosphere. Some scientists say they have detected some gases existing around Pluto.

ASTEROIDS

Asteroids are sometimes referred to as minor planets. The largest number of asteroids in our solar system are in the space between Mars and Jupiter. This area is called the asteroid belt. More than 2,000 asteroids have been charted in this area, although not all have been named. Many of these asteroids are only a foot or so in diameter.

Ceres is by far the largest asteroid; Pallas is the next largest. Asteroids are basically large or small chunks of rock. Their orbits around the sun can range from elliptical to any variation on a circle, and very few are actually round. It is not unusual for asteroids to collide, causing them to change shape and/or orbit.

The larger asteroids will appear as disks through larger telescopes. The diameters of Ceres, Pallas, and other larger minor asteroids have been determined in this manner. Asteroids may rotate on an axis in the same way that planets do.

The orbits of some asteroids intersect Earth's orbit occasionally. These asteroids have been dubbed Apollo objects, because Apollo was the first to cross Earth's orbit. The odds of an asteroid actually hitting the earth are very slim.

Asteroids have little, if any, gravitational pull and no atmosphere. Refer to an astronomical almanac or guide to obtain their current positions. Many larger asteroids will be easily viewed with a small telescope.

COMETS

Although no two *comets* are exactly alike in composition or orbit, each is composed of three basic parts: the *nucleus*, which is in the center and is quite bright and very small; the *coma*, which is a much fainter and nebulous structure surrounding the nucleus; and the *tail*. The nucleus and the coma form the head of the comet.

The tail of a comet is formed in direct proportion to its proximity to the sun. Some orbits of comets are elliptical, while others are quite erratic. Regardless of the exact path, the majority of comets reside in a belt in space beyond Pluto's orbit. As the comet's orbital path brings it nearer to the sun, it begins to change. Because comets are generally located in the outer regions of outer space, they have a temperature close to absolute zero. Most astronomers theorize that the outer layers consist of frozen methane, ammonia, and water mixed with what appear to be stony particles. As the comet leaves the outer reaches beyond Pluto and approaches the sun, it begins to melt, forming a gaseous tail. The tail points away from the sun, as it is affected by the solar winds.

A comet's orbit may be affected by the gravitational pulls of planets that the comet nears during its travels. Jupiter has captured many comets and changed their orbits considerably.

The size of a comet and its tail is an endless source of amazement. Halley's comet is reported to have been measured as approximately 94 million miles. The largest comet measured to date is an impressive 200 million miles. These dimensions do not remain constant and will be greatest at the point when the comet is nearest to the sun, or at *perihe-*

lion. As the comet travels away from the sun, the tail begins to shrink as the particles and gases begin to solidify. Some of this material is lost to space.

A spectroscope is used to study the composition of comets. Scientists have been able to determine that comets are composed of nickel, silicon, manganese, calcium, and neutral iron.

Obtain a schedule as to when those comets already identified are scheduled to appear again. There are usually about 20 known comets that can be viewed each year. The best time to search for comets is shortly after dark in the western sky and shortly before dawn in the eastern sky. The brightest comets will be found close to the sun when they are at perihelion.

METEORS AND METEORITES

A *meteor* is basically a chunk of rock or metal that is usually the result of a disintegrating comet after it has passed perihelion, and pieces of its tail begin to fall free. A *meteorite* is a meteor that has survived the heat of earth's atmosphere and lands on its surface. A meteor is the piece of rock that has become heated and appears as a "shooting star" in our skies. Meteors can also be pieces of asteroids.

Meteorites have been divided into three basic categories regarding composition: stony iron, iron, and stony. Those meteorites composed of stone are the most common and also the largest. These meteorites are quite similar to a type of silicate rock found on earth. Iron meteorites are actually a mixture of both iron and nickel. They are found as frequently as are stony meteorites. The stony iron meteorites are composed of stone and metal and are not found as frequently.

A meteor shower is usually the result of a disintegrating comet whose orbit the earth has intersected. The particles of the comet often tend to move parallel to one another almost in the same path and direction as the original comet. The number of these meteor showers is related to the seasonal changes on earth. Meteor activity increases during the night, with the highest amount of activity occurring in the early morning hours. Those meteors seen during the peak period will also be the brightest. Observing a falling star or meteor shower with the naked eye is always exciting. If you view a meteor shower through a telescope, the effect will be brilliant.

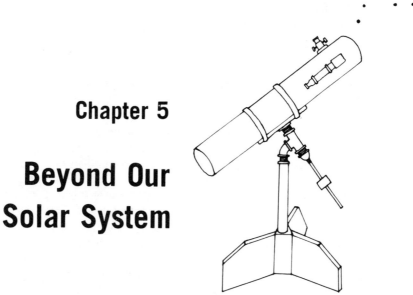

Chapter 5

Beyond Our Solar System

WHEN ASTRONOMERS BEGAN TO LEARN more about deep space, they were truly amazed amazed at the vastness of our galaxy and those beyond it. Astronomers determined the location of the sun and planets in relation to our Milky Way galaxy in this century. These discoveries can be directly attributed to modern advances in sky-viewing instruments.

Many names given to stars and constellations (groups of stars) date back to early civilizations. A majority of stars and constellations named during ancient times carry the names of animals. Astronomers today may find this confusing, because few stars or constellations resemble animals after which they are named.

Professional observers have developed mathematical systems and identification methods with which to aid the amateur in studying deep space. Maps complete with pictorial views of all sections

of the skies will prove invaluable in observing deep space.

STARS

The sun is the nearest star to earth and is the center of our solar system. When astronomers realized the immense distances separating earth from the other stars, they devised a system where distance is measured in time—hence, the term light year. The speed of light is approximately 186,000 miles per second (light second). The distance traveled in one year would be a multiple of 186,000 expanded to obtain a figure of 5.88 billion miles in one light year. The sun is approximately 93 million miles from earth, or 8 light minutes if converted to the light-year form of measurement.

Determining stellar distances was the first and probably most significant step in the study of stars.

When the distance that light must travel to reach earth is known, the magnitude of each individual star can be accurately charted. Without knowing the distance, it might be assumed incorrectly that two stars appearing to be of equal brightness are of equal magnitude. The system of measuring each star's light is called *apparent magnitude* and is based on a numerical system. This method has been around since the early civilizations when observations were based solely on brightness. The brightest stars were of the first magnitude, those less bright were of the second magnitude, and so on. Only those stars visible to the naked eye were labeled. This system has been adjusted. Some stars are even brighter than those originally determined to be of the first magnitude. Rather than completely do away with this system, astronomers found it simpler to devise a mathematical system. They were able to determine that the stars of one magnitude were separated from the next lower magnitude by a 2.5 factor. A star with a magnitude of 3.0 is 2.5 times brighter than a star with a magnitude of 4.0. This method has greatly simplified observations and studies of stars.

A star is formed in those areas in deep space where there is a huge cloud of dust and gases known as a *nebula*. As the mass grows in size and density, it gathers up additional dust and gas by gravitational force, causing the inner temperatures and pressure to increase. A star is really a ball of gas that is mostly hydrogen. The star is quite cool in these early developmental stages. As temperatures increase, a thermonuclear reaction occurs. Hydrogen is converted into helium, at which point the mass is so hot that it shines. It is called a main sequence star and will probably remain so, much like the sun, for many thousands of years. A lack of hydrogen eventually ends the life of a star. When the hydrogen has been completely converted to helium, the thermonuclear reaction ends and the star begins to cool. The star's own gravitational force begins to shrink, pulling the matter in closer and closer. The temper-

ature inside the core of the star increases, and the mass expands to a size much larger than its size at the point where it was classified as a main sequence star. A star in this stage will appear through a telescope as a red mass of light of varying intensity and brightness. These stars are called red giants.

The sun is presently a main sequence star and will probably be in this stage for another 500 million or more years. A star that is a red giant may remain one for just as long. When a star becomes a red giant, it is not considered a stable star. Stars that pass through the main sequence stage due to the reactions of their matter become nonstable or variable stars. The main difference between a stable and unstable star is that the energy being released in a stable star remains at a fairly constant rate. The star's brightness or luminosity remains the same. A star becomes more unstable as hydrogen is converted to helium, and it cools down.

The next stage in a star's life will depend on the size of its mass. Extremely large stars will go through further processes until their inner core is composed of carbon and other elements. The heat will become so intense that this huge mass will eventually explode. This is called a supernova. The novae are more common and are interesting to amateur astronomers. A nova is caused by the explosion of a less massive star. It is usually quite noticeable because a star will suddenly increase in brightness and intensity

After completing the red giant stage, a smaller star will cool down and shrink to the point where it is merely a faint white light. These stars are white dwarfs and are quite dense. A white dwarf is one of the late stages in a star's life. This stage may last for billions of years, until the star has cooled to the point where it eventually dims and goes out completely.

Cepheids are stars that vary in brightness over a specific period in a somewhat consistent manner. Astronomers believe these stars vary in luminosity due to the actions occurring at their core as a result of the thermonuclear process. One such star that is clearly visible with the naked eye is Delta Cephei, which almost doubles in brightness during the stages that it varies in luminosity. The time period involved is usually quite long. The period of variability is directly related to the star's true luminosity.

Another class of variable stars is the RR Lyrae group. These are usually less consistent regarding brightness and have shorter time periods during which their variations occur. They are most often found in globular clusters and will appear as either white or yellow-white spots of light when viewed through a telescope. The cause for these variations is believed to be related to the thermonuclear reaction occurring at the star's core.

Planetary nebulae are stars that have nothing to do with the planets. When viewed through a telescope, these stars appear in disk-like shapes of planets. Planetary nebulae may be the result of gas explosions from other stars already discussed. If the idea is accurate, this is yet another stage in a star's life.

CONSTELLATIONS AND DOUBLE STARS

An atlas will help you locate constellations. There are 88 constellations at this writing. Although many of the constellations have descriptive names, these descriptions do not match what you will observe through the telescope. There is an ever-changing movement occurring among the stars, although this movement is basically imperceptible because of the vast distance involved.

There is no real scientific significance involved in the identification process. The locations and brightness of different stars in these constellations are very useful to the astronomer as guides for further viewing, and they play a large part in navigational operations of aircraft.

Purchase some star charts and constellation maps. These are available in many bookstores. Some mail-order companies that offer a complete line of telescopes have these charts and maps at very reasonable rates.

Orion is probably one of the more popular constellations among amateur and professional astronomers. It consists of many bright stars in varying stages of their life cycles. Orion is visible during the winter months. One of its most striking stars is Betelgeuse which is a massive, brilliant red giant. Orion's name comes from early Greek mythology. It represents a hunter with a sword and shield. This constellation contains many gas clouds or nebulae. The Great Nebula will actually appear to glow when viewed through a telescope. Astronomers have been able to observe the formation of new stars in the Great Nebula. Orion also contains another nebula known as the Horsehead. It may be more difficult to observe through the average telescope due to its distance.

Some double stars will be clearly visible with the naked eye. Many double stars will appear as a single star at first. When observing them more closely, usually through a spectroscope, they can be split and seen as two separate masses. This phenomena is not all that unusual, because two, three, or even several stars can travel together with a common center of gravity. One star in a double star configuration may be very big, and the other may be very small. The two stars are probably the same age.

Stars in double star configurations can be classified into *optical* and *binary* stars. Binary stars actually maintain a similar orbit and possibly exchange gases. Optical stars are really an illusion caused by the fact that two stars may be directly in line with each other, but they are actually separated by a lot of space. Binary stars are usually not visible except through high-powered observatory telescopes, although some are close enough to be viewed through quality telescopes.

One of the binary double star configurations that can be viewed is in Orion. There are several double stars in Orion. The star Rigel has a smaller companion that, in turn, has two smaller companion stars.

STAR CLUSTERS

A *star cluster* is a group of stars in a huge mass located in our galaxy. Star clusters fall into two general categories: *open star* clusters and *globular star* clusters. Open clusters are loosely arranged groups of stars. They often are not too distinctive from the background stars. These clusters can best be seen through low-power, wide-field oculars from dark sky locations. Globular star clusters are tightly packed, spherically-shaped groups of thousands of stars. Moderate to high power will show these objects best.

Open Star Clusters

Open star clusters are generally located in a line with the plane of the Milky Way. This arrangement makes it difficult to identify them through a telescope, because they tend to mix with other background stars. A common gravitational force holds this large group of stars together. It is commonly thought that the group originated at the same time, with the differences in the varying stages of the stars being caused by the variation in size.

One of the most popular eye-appealing open star clusters is the Pleiades cluster. It consists of approximately 300 stars in varying stages and presents a colorful view through a telescope. Other open star clusters that can be viewed through a telescope are M35, in the constellation Gemini, and M11, which is visible in the constellation Scutum during the summer.

Globular Star Clusters

Globular clusters are much older than open clusters. A globular cluster can consist of hundreds of thousands of stars. They are usually located near what is considered to be the center of our gal-

axy—at a great distance from the sun.

The age of most globular clusters is known due to intensive study of the brighter stars. There are numerous red giants and even planetary nebulae. Some of the stars in these particular globular clusters have thus exploded.

Globular clusters are found in a closely knit group in the constellation *Sagittarius*. The remaining known globular clusters are spread out over the outer edges of the galaxy and form a sort of halo around the Milky Way. Figure 5-1 shows globular cluster M13 in Hercules. It is generally regarded as the finest globular cluster in the northern sky.

Fig. 5-1. Globular cluster M13 in Hercules (courtesy Celestron International, Torrance, California, U.S.A.).

NEBULAE

There are planetary and diffuse nebulae. Most

planetary nebulae shine with a greenish glow and have a round or elliptical shape. Planetary nebulae are best seen when using moderate to high magnification from dark sky locations. To see some really faint details, try averting your vision. Glance off to the side of the field of view rather than looking directly at the object of interest.

The Ring nebula (M57) is one of the loveliest planetary nebulae. Resembling a nearly perfect smoke ring, the various contrast levels of the nebulosity are evident. When viewing this particular nebula, look closely for the dull illumination in the ring's center. The extremely faint central star may be occasionally glimpsed. The Ring nebula is visible during the summer months in the constellation Lyra (Fig. 5-2).

The Dumbbell nebula (M27) has scores of stars apparently embedded in its oval (Fig. 5-3). The Dumbbell is also a summer object visible in the constellation Bulpecula.

Diffuse or emission nebula are larger and brighter than planetary nebula. The word "bright" sometimes gives an exaggerated notion of the amount of light given off by these nebulae. The published magnitudes of nebulae are the magnitudes these objects would have if their images were compressed into the size of a single stellar image. Do not expect a third magnitude nebulae to appear as bright as a third magnitude star.

Diffuse nebulae are vast, irregularly shaped clouds of rarefied gas. They are spurred into luminescence by radiation from nearby stars, or because they reflect the light of nearby stars. These large, dim objects are best seen using low-power, wide-field oculars. Observe these objects from a dark site and use averted vision to see the really faint details.

The Great Nebula in the winter constellation Orion (M42) is one of the most magnificent sights in the sky. It will fill a low power field of view with its intricate, filamentary network, which is laced with many knotty brightenings. The dark clouds of the nebula are crisply defined.

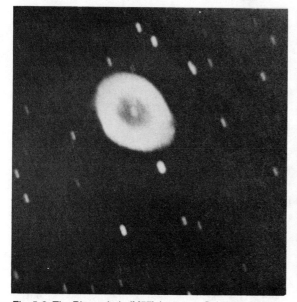

Fig. 5-2. The Ring nebula (M57) (courtesy Celestron International, Torrance, California, U.S.A.).

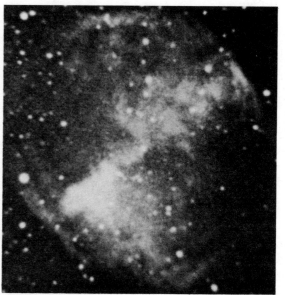

Fig. 5-3. The Dumbbell nebula (M27) (courtesy Celestron International, Torrance, California, U.S.A.).

The Lagoon nebula (M8) is one of the largest, star-studded wonders in the summer Milky Way. It consists of many beautiful shades of colors. Its swirling nebulosity is divided nearly in half by a huge, dark lane (Fig. 5-4). An open star cluster, NGC 6530, is embedded in the nebula.

the sky. These nebulae are composed of gas and dust, but they are not close to the bright stars that make the bright nebulae appear so luminous and brilliant. Sometimes bright and dark nebulae are located near each other, and both can be observed at the same time.

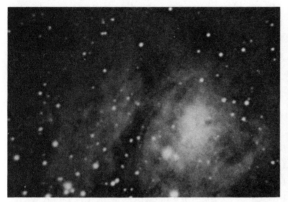

Fig. 5-4. The Lagoon nebula (M8) (courtesy Celestron International, Torrance, California, U.S.A.).

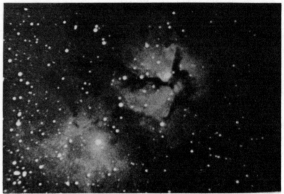

Fig. 5-5. The Trifid nebula (M20) (courtesy Celestron International, Torrance, California, U.S.A.).

Slightly north of the Lagoon is the Trifid nebula (M20) (Fig. 5-5). Both components are well-detailed, with the brighter part clearly trisected by dark lines. Both the Trifid and Lagoon nebulae are located in the constellation Sagittarius.

Diffuse nebulae are most often found in the Milky Way, and they are sometimes referred to as galactic nebulae. They are usually located near an area of populated bright stars, which accounts for their brightness. Other elements must comprise the nebula besides its main component, hydrogen, because gas by itself would not be able to reflect light. Cosmic dust particles give off their own fluorescence that contributes to the brightness of nebulae.

Diffuse nebulae may be bright or dark. These bright nebulae are aided in their brightness by nearby bright stars, as well as by the luminous cosmic dust of which they are partially composed. Dark nebulae have become apparent to astronomers partly because they have blocked objects in

Dark nebulae will often appear as *black holes* or voids because of this lack of luminosity. Astronomers and scientists are unable to accurately determine the distances of these nebulae. Much study has gone into attempting to calculate the distances, and rough estimates have been made by calculating the distances of those stars nearest to the nebulae.

Many theories have been presented as to the origin of diffuse nebulae. The most popular theory is that they are somehow associated with the formation of stars. This theory is based on the fact that the nebulae are composed of hydrogen, helium, and oxygen—the same elements as stars. A star is formed by a dense cloud of these materials gathering together and forming as a result of gravitational force. This same occurrence can be loosely applied to the formation of nebulae. Nebulae are not sufficiently dense, and the pressure remains much lower. The thermonuclear action associated with stars does not occur.

GALAXIES

Advanced studies of nebulae over the years brought about the discovery that some nebulae thought to be in our galaxy were actually far beyond those dimensions. This discovery was greatly hampered by the clouds of dust and gas located along our own galaxy which hid from view what lay beyond it. With the advent of higher-powered observational instruments, astronomers discovered that these nebulae were really galaxies composed of many billions of stars. Some of those objects viewed were diffuse nebulae. These have been termed extra galactic nebulae to differentiate between those in our galaxy and those outside it.

Astronomers have been able to somewhat classify these universes into three basic categories according to their shape, size, and structure. Galaxies exist in a variety of sizes and regular and irregular shapes, and most are faint. They are best seen from a dark sky location. Large galaxies, or clusters of galaxies, are best seen when using low-power, wide-field oculars. Smaller galaxies are better seen when using moderate magnification.

The classification of galaxies was devised by Edwin Hubble in the 1920s. His system divided the galaxies into three basic groups: *elliptical, spiral*, and *irregular.* Many modifications to these classifications have been made.

Elliptical Galaxies

Galaxies that present an elliptical appearance do so primarily because of the speed at which they rotate. Galaxies in this category do not all appear in the same elliptical shape. They have been further subdivided regarding their individual amount of ellipticity. Some of the larger stars in the closest elliptical galaxies are composed of some red giants, RR Lyrae stars, and a pronounced lack of dusts and gases. Elliptical galaxies, usually the largest of the three different types, are somewhat transparent. It is possible to "look through" an elliptical galaxy and see other galaxies beyond. Astronomers believe that the elliptical galaxies are the oldest and the largest of the different types observed thus far. These island universes may contain planets.

Most galaxies that have been observed and charted are smaller than the Milky Way. Astronomers believe that those galaxies located the farthest away from earth are moving at greater velocites. Studies have indicated that they are moving on a path away from us at this high speed. The universe may be going through a continuous, high speed expansion process that may eventually reverse itself.

Spiral Galaxies

Spiral galaxies are the most prominent and comprise approximately 80 percent of those galaxies known today. Probably the most well-known and easiest to observe of the spiral galaxies is the Andromeda galaxy (Fig. 5-6). Its core is ablaze with the light of possibly, a million suns. You should easily observe the dark lanes separating the faint, outer spiral arms, as well as emission nebulae within the arms.

The core of a spiral galaxy has been perceived as similar to that of an elliptical galaxy. A spiral

Fig. 5-6. The Andromeda galaxy is well-known among amateur astronomers (courtesy Celestron International, Torrance, California, U.S.A.).

galaxy is composed of many "arms" consisting of numerous bright stars. These spiral arms are also composed of nebulae, star clusters, and gaseous substances. The Andromeda galaxy and other galaxies have companion galaxies nearby.

Irregular Galaxies

Irregular galaxies have no particular shape and no central core. An irregular galaxy is the Magellanic Clouds, which have no particular shape. It is theorized that an irregular galaxy is actually in the beginning stages of its evolution, and it may possibly develop over an extremely long time into a spiral galaxy. This theory has been introduced due to intensive observations indicating that the Magellanic Clouds have some characteristics of spiral galaxies, such as very faint spiral arms. These arms may possibly appear so faint because they are in the very early evolutionary process.

In irregular and spiral galaxies there appears to be a continued evolution occurring in the form of new stars, varying formations of gas clouds, etc. This has not been the case with elliptical galaxies. They may be older and farther along in their evolutionary process. All galaxies may start out as irregular ones and progress over billions of years into eventual elliptical galaxies of immense size. This is only one theory, and many have argued that this is not possible or even logical.

Our Milky Way and the Andromeda galaxy have a number of companion galaxies. Several very irregular galaxies are connected by an immense field of stars. The Whirlpool galaxy shown in Fig. 5-7 is an example of this feature. The magnetic pull of each separate galaxy may cause them to exchange materials. Even more interesting is the observation of two or more galaxies that appear to have merged. Sometimes a galaxy will actually shoot out a stream of light into its surrounding space. These streams are sometimes even larger than the length of the galaxy.

Fig. 5-7. The Whirlpool galaxy is composed of a number of interconnected galaxies (courtesy Celestron International, Torrance, CAlifornia, U.S.A.).

Theories have been introduced to explain these unusual phenomena. One idea is that as an individual star in the galaxy explodes, it may create a chain reaction effect. Other stars may do the same, resulting in a massive burst. Another idea is that as the magnetic forces toward the galaxy's core build up to an intense point, the result is a reaction that causes the explosion. This may also be caused by a very thick density of stars, gases, and other matter.

Much of the identification process involving all galaxies, and irregular galaxies in particular, has been attributed to the extremely strong radio waves they emit. This may be due to the extremely intense action occurring in irregular galaxies. All galaxies do emit radio waves, but some are much stronger and travel greater distances.

Because astronomers have learned that observable galaxies are moving away from us at great velocities, it is logical to assume that there are countless more galaxies at distances so vast that they cannot be observed with existing telescopes.

Quasars are tiny points of light that travel away from us at an extremely high rate of speed and produce amazing amounts of energy. Very little is known about these objects because of their great distance from earth.

Dwarf Galaxies

Our own galaxy is surrounded by many small dwarf galaxies. These galaxies went unnoticed for a time and were thought to be simply clouds of gas and dust particles. They are also commonly found in other clusters of galaxies and superclusters. A dwarf galaxy may be difficult to observe through even a quality telescope. Most dwarf galaxies have been discovered and charted by large observatories equipped with giant telescopes.

Chapter 6

Using the Telescope

WHEN YOUR TELESCOPE HAS BEEN SET UP, YOU can use it to make observations on land and in the skies. Most manufacturers will provide a detailed instruction manual that tells how to set the telescope up and describes the controls.

A telescope is a precise instrument that has been aligned at the factory. You should have no difficulties operating the telescope when you are familiar with its functions and capabilities. Proceed slowly and patiently. Start with the simplest techniques and move to more technical procedures. Note the placement of all accessories and controls.

The telescope should be mounted in a stable and secure position, particularly for long observing periods. If your telescope is a small, portable instrument, it will be easy to provide this stability. Larger telescopes are usually equipped with some type of mounting or stand that will provide a secure base for the instrument.

Practice using the telescope for terrestrial observing before pointing it upward. This will familiarize you with focusing, the different eyepieces, and other components such as the alignment and star diagonal. Even though the image will not be erect, you will gain valuable experience in handling the telescope.

Allow sufficient time for the telescope to adjust to the outdoor temperature. If the outdoor temperature is very low and the instrument has been stored in a heated building, the cooling period may take up to 1 hour. The telescope may be used during this period, but satisfactory results will be obtained only at low power.

OBSERVING AT NIGHT

When the telescope has adjusted to the temperature, you are ready for some night observations. Point the telescope at a bright sky object—

is fairly bright and easy to find—such as the moon or a bright star. Look straight into the eyepiece, keeping just outside of eyelash range. You should see some light. If not, move your head slowly above the eyepiece until you do. It takes your eye some time to adjust to darkness. If the temperature outdoors is cool, stay indoors with your eyes closed or sit in a dark room for a short period until your eyes adjust to the darkness. If the telescope has already been set up outdoors, it is going through its own adjustment. If you plan to be using any maps, charts, or notes while outdoors, use a lamp or flashlight covered with red or brown paper, or possibly a red filter.

Your eye must not touch the eyepiece, but it must be centered on the emergent light beam. This will be difficult to do if your eyes have not adjusted to the darkness. Note that the sky as seen through the telescope is not really black but rather a bright, luminous shade of gray. Given this target, your eye will automatically center on the eyepiece. You can cup a hand around the eyepiece to act as a guide until your eye is centered on the light beam.

If you happen to wear eyeglasses, take them off if you are farsighted. You will then see distant objects clearly. The removal of the glasses will let you crowd the eyepiece when necessary. If you are nearsighted, you have a different problem. If you remove your glasses, you cannot see distant objects. Keep the glasses on and use only eyepieces with long eye relief of ½ inch or more. You should also keep your eyeglasses on if you have astigmatism.

Avoid sighting through mist, fog, haze, or heat waves. No telescope can cut through these obstructions. Seek out dark, steady skies for nighttime celestial observing. Very dark skies are best for nebulae and galaxies, and very steady skies are best for the moon and planets. If you find a dim nebula difficult to see, try averting your vision (glancing to the side in your field of view) or moving the field of view back and forth slightly to bring the more sensitive outer portion of your retina into use.

Viewing is good when atmospheric turbulence is at a minimum. You can determine this with the naked eye by observing how much the stars appear to twinkle. When the stars shine with a steady glow rather than twinkle, the viewing is steady. Deep sky observing of nebulae and galaxies is not nearly as affected by seeing conditions as lunar and planetary viewing. The most important factors are the transparency of the atmosphere and the darkness of your observing site. The advantages of observing deep sky objects from a dark sky location cannot be over-emphasized. From a dark sky site, you will see the faint, filamentary details usually seen only in observatory photographs.

FOCUSING

When light is visible in the eyepiece, turn the focusing knobs to clarify the image. There is no such thing as exact focusing of a telescope. The image forms at a very precise and exact image plane, but you can see the image at various settings of the eyepiece because the eye can adjust for either long or short focus. The best practice is to focus long. This is done by extending the eyepiece a little more than necessary and then focusing in just enough to get a sharp image. The long focus causes the eye to focus as for a distant object—the most comfortable position. If you focus to the maximum in position that retains a sharp image, the eye accommodates for a close object. This position gives slightly greater magnification but is somewhat more tiring during extended use. You should use both the long and short focus, because frequent changes will allow you to see clearer without eye fatigue.

If you are looking at a star, turn the focusing knobs until the light becomes a clearly defined point. Then note that other points of light will suddenly appear. These are the fainter stars that become visible as you resolve the target image. Don't be surprised when you see the object upside down through the telescope. The image is inverted in virtually all astronomical telescopes.

MAGNIFICATION

The utility of any given magnification will depend upon the object's apparent size, its apparent brightness, and the seeing conditions. High powers tend to decrease image brightness, diminish the field of view, and magnify air turbulence. Planets, lunar craters, some globular clusters, and planetary and diffuse nebulae will be most pleasing at magnifications ranging from 100 × to 220 ×. Go to higher power if you want to observe the moon and planets in greater detail. Magnifications between 220 × and 435 × will be very useful.

Extremely high magnification oculars can be profitably used on objects of suitable brightness and on nights of extremely fine viewing. Lunar and planetary detail and close double stars are suitable subjects. Be sure that the atmosphere is stable enough for such magnifications. Pronounced star twinkling indicates turbulence overhead. High power will simply magnify the effects of such turbulence, while the lower powers may produce a steadier and more satisfactory image.

The main body of the earth's atmosphere is about 10 miles thick straight up. At 45 degrees it is about 15 miles thick, while the air blanket can be 100 miles or more near the horizon. The best seeing is at the zenith where the air blanket is thin. The atmosphere is constantly in motion—shifting, swirling, and boiling. It is a rare night when you can use powers more than 300 ×, regardless of the size or excellence of the telescope. Atmospheric disturbance will seldom be a problem at 50 × to 100 ×, except when you are attempting to observe through an open window or over a hot chimney. Window viewing is practical when both the indoor and outdoor air are at about the same temperature. A power of 50 × will be about the maximum for this type of viewing.

DECLINATION SETTING CIRCLES

The telescope's *declination setting circles* should be aligned so that the 90-degree-90-degree line on each parallels your telescope's optical axis. When the optical axis is parallel to the polar axis, the declination pointer on the fork tine at the bottom of the declination circle should give a reading of 90-degrees. (The more specific information here applies to Celestron products.)

To set the circles accurately, first orient your telescope tube with the finderscope up. Center a star or a planet in the field of your main optics. Note the declination reading on one of the circles. Tumble the telescope tube in both right ascension and declination until the finder is under the tube, and you have the same star centered in the field again. Note the declination reading (on the same circle). It should be the same as before. If the reading is not the same, rotate the circle back to its proper position. The correct position will be such that the coordinate exactly halfway between your first and second readings is opposite the declination pointer. For greatest accuracy, repeat this procedure until the identical reading is obtained after the tube is tumbled. This will also be the correct reading for the other declination circle.

CELESTIAL-COORDINATE SYSTEM

The *celestial-coordinate system* is an imaginary projection of the earth's geographical coordinate system onto the starry sphere that seems to turn overhead at night. This celestial grid is complete with equator, latitudes, longitudes, and poles. It remains fixed with respect to the stars. The celestial-coordinate system is actually being displaced at a very slow rate with respect to the stars, because the earth's axis is very slowly changing in the direction of its point. This effect is very slight.

The celestial equator is a full 360-degree circle bisecting the celestial sphere into the Northern Celestial Hemisphere and the Southern Celestial

Hemisphere. Like the earth's equator, it is the prime parallel of latitude and is designated 0 degrees. The celestial equator passes through the constellations Orion, Aquila, Virgo, and Hydra.

The celestial parallels of latitude are called coordinates of declination. Like the earth's latitudes, they are named for their angular distance from the equator. These distances are measured in degrees, minutes, and seconds of arc. Declinations north of the celestial equator are +, and declinations south are −. The poles are at 90 degrees.

The celestial parallels of longitudes are called coordinates of right ascension. Like the earth's longitudes, they extend from pole to pole. There are 24 major right ascension coordinates evenly spaced around the equator—one every 15 degrees. Right ascension coordinates are a measure of time and angular distance. The earth's major longitudes are separated by 15 degrees, but they are also separated by one hour of time, because the earth rotates every 24 hours. The same principle applies to celestial longitudes, because the celestial sphere appears to rotate once every 24 hours.

Astronomers generally prefer the time designation for right ascension coordinates, even though the coordinates denote locations on the celestial sphere. It is easier to tell how long it will be before a star will cross a particular north-south line in the sky. Right ascension coordinates are marked off in units of time eastward from an arbitrary point in the constellation Pisces. The prime right ascension coordinate that passes through this point is designated "0 hours 0 minutes 0 seconds." All other coordinates are named for the number of hours, minutes, and seconds that they lag behind this coordinate after it passes overhead moving westward.

It is possible to find celestial objects by translating their celestial coordinates into telescope point. Celestron and other telescopes are equipped with setting circles. The exact location will vary. The dial at the base of the Celestron telescope is the setting circle for right ascension (Fig. 6-1). The

R.A. SETTING CIRCLE

Fig. 6-1. The dial at the base of the Celestron telescope is the setting circle for right ascension (courtesy Celestron International, Torrance, California, U.S.A.).

dials at the top of the fork tines are the setting circles for declination. You can use these circles to acquire celestial objects when you have properly mounted the telescope and pointed the polar axis of the instrument toward the North Celestial Pole.

LINING UP ON THE POLE

The *celestial pole* is that imaginary point on the celestial sphere toward which the earth's axis of rotation points. Stars appear to move nightly around this point. Their paths are concentric circles with the celestial pole at the center. If the polar axis of your telescope points directly at the celestial pole, a star at any declination may be kept centered in the telescope's field simply by rotating the instrument in right ascension, or by letting the electric clock drive of your telescope rotate for you in right ascension.

A simple polar alignment on the north star, Polaris, is adequate for casual viewing. Polaris, which is within 1 degree of the true North Celestial Pole, is easy to find. The pointer stars in the bowl of the Big Dipper point straight to Polaris.

Tilt the telescope tube until the declination circle reads 90 degrees. Move the tripod and adjust the wedge until Polaris is in the center of the field of view. The telescope is now ready to be used. The circles will read to within approximately 1 degree accuracy, and the drive will keep an object in the field of view for a considerable period.

To achieve a more accurate polar alignment after aligning on Polaris, repoint the telescope at a bright star near the celestial equator. Look up that star's right ascension in a star atlas. Move the right ascension setting circle until the right ascension pointer is indicating that right ascension. Turn the telescope in right ascension until it indicates the right ascension of Polaris. Lock the right ascension clamps. Move the tube only in declination until the declination pointer indicates 90-degrees. Continue moving the tube in the direction away from the Big Dipper—toward Cassiopeia—until the declination reads +89.2 degrees (the declination of Polaris). Lock the declination clamp. Move the tripod and adjust the wedge until Polaris is centered in the field of view.

PRECISE POLAR ALIGNMENT FOR ASTROPHOTOGRAPHY

The telescope is now aligned well enough for you to try deep sky photography. There are several advantages in precisely aligning your telescope to the true north pole. There will be no image drift in declination with an exact polar alignment. Also, there will be no star trailing caused by field rotation, the tracking will be more accurate, and your setting circles will read very accurately. Because the exact polar alignment eliminates the need to make corrections in declination during long exposure astrophotography, it allows you to concentrate on right ascension corrections.

After the quick alignment methods described previously, you will need an illuminated reticle eyepiece for this more precise method. A Barlow lens will also speed the procedure considerably. Insert the illuminated reticle (and Barlow lens, if used) and repoint the telescope at a fairly bright star near where the meridian and the celestial equator intersect (preferably within ±½ hour right ascension of the meridian ±5 degrees of the celestial equator). Monitor the declination drift (ignore any drift in right ascension). If the star drifts south, the polar axis points too far east. If the star drifts north, the polar axis points too far west.

Move the telescope's polar axis in the appropriate direction until the north or south drift stops. Accuracy of this adjustment will be increased if you use the highest possible magnification and allow the telescope to track for a period.

Repoint the telescope at a fairly bright star near the eastern horizon and near the celestial equator (the star should be at least 20 degrees above the horizon and ±5 degrees from the celestial equator). If the star drifts south, the polar axis points too low. If the star drifts north, the polar axis points too high.

Monitor only the declination drift using high magnification for a period. After you have made the adjustments to stop the declination drift, you will have achieved a highly accurate polar alignment. This same procedure may also be employed by Southern Hemisphere observers, but the directions of drift will be reversed.

USING THE SETTING CIRCLES

The right ascension setting circle on most

telescopes is usually marked with divisions representing 5 minutes each. Regarding a declination setting circle, each graduation usually represents 1 degree. The exact location and designations may be different than those described here for the Celestron 14 telescope.

When you have lined up on the pole and set your right ascension setting circle, you can use the setting circle readings to translate the star atlas coordinates of a celestial object into telescope point. To set the right ascension setting circle, center a star of known right ascension in your telescope's field. To aid you in selecting a star, refer to the alphabetical listing of stars in Table 6-1 and the list of Messier objects in Table 6-2. Rotate the circle (it will turn freely) until the coordinate of the star is under one of the two right ascension pointers. Use the right ascension pointer that is most convenient to see. After setting the right ascension with one pointer, remember that the other pointer will read 12 hours off. If you switch from one right ascension pointer to the other, allow for this

incorrect reading (or reset the circle using the other pointer).

Now that the right ascension circle is set, use a star atlas to look up the coordinates of the objects you wish to observe. Rotate the fork mounting until the right ascension of the object you selected is indicated. Lock the right ascension clamps. Move the tube in declination until the proper declination is indicated, then lock the clamp.

Use your lowest power eyepiece when trying to locate a celestial object. If you don't find the object quickly, check your finderscope. Most objects will be visible through it, and you can quickly center the object. If your finder is accurately aligned, the object will then be in the main telescope's field. When you want to seek out another object, release the clamps and move the telescope until the proper coordinates are indicated.

LUNAR AND PLANETARY PHOTOGRAPHY

Although Cassegrain-focus photography is excellent for small-scale renderings of the moon and planets, extremely long focal lengths are necessary to photograph finer details and to get a reasonably large planetary image on film. A method of eyepiece projection is needed. Celestron offers a device known as the Tele-Extender. The ocular acts as an enlarging lens, projecting a magnified Cassegrain-focus image onto the film. The effective focal length may be varied by replacing the ocular with one of a different focal length. The 25-mm, 18-mm, and 12-mm are the most useful oculars for eyepiece projection.

When the Celestron Tele-Extender is properly attached, successful high magnification photography will still be extremely dependent upon steady viewing conditions. Shoot only when the viewing is steady. Focus very carefully. If the object is too dim for easy focusing, try focusing on a bright, nearby star and then move the telescope back to the desired object. If your camera has interchangeable focusing screens, change to a perfectly clear (aerial

Table 6-1. Alphabetical Listing of Stars (courtesy Celestron International, Torrance, California, U.S.A.).

Star	Constellation	Apparent Magnitude	1970 Position R.A. (h/m)	1970 Position Dec. (°/')
Achernar	Eridanus	0.6	0137	−5724
Acrux	Crux	1.4	1225	−6259
Aldebaran	Taurus	1.1	0434	+1627
Altair	Aquila	0.9	1949	+0847
Antares	Scorpius	1.2	1628	−2622
Arcturus	Bootes	0.2	1414	+1921
Bellatrix	Orion	1.7	0524	+0619
Betelgeuse	Orion	0.1	0554	+0724
Canopus	Carina	−0.9	0623	−5241
Capella	Auriga	0.2	0514	+4558
Deneb	Cygnus	1.3	2040	+4510
Fomalhaut	Piscis Aust.	1.3	2256	−2947
Pollux	Gemini	1.2	0743	+2805
Procyon	Canis Minor	0.5	0738	+0518
Regulus	Leo	1.3	1007	+1208
Rigel	Orion	0.3	0513	−0814
Sirius	Canis Major	−1.6	0644	−1640
Spica	Virgo	1.2	1323	−1100
Rigil Kent	Centaurus	0.1	1438	−6043
Tureis	Carina	2.2	0916	−5909
Vega	Lyra	0.1	1836	+3836

Table 6-2. List of Messier Objects (courtesy Celestron International, Torrance, California, U.S.A.).

Desig-nation	1970 Coordinates R.A. (h/m)	Dec. (°/′)	Con.	Mag.	Type Object	Comments
M1	0533	+2200	Tau	8	P. Neb.	*
M2	2132	−0058	Aqr	6	Gl. Cl.	
M3	1341	+2832	CVn	6	Gl. Cl.	
M4	1622	−2627	Sco	6	Gl. Cl.	
M5	1517	+0212	Ser C	6	Gl. Cl.	
M6	1738	−3212	Sco	5	Op. Cl.	
M7	1752	−3448	Sco	5	Op. Cl.	
M8	1802	−2420	Sgr	7	D. Neb.	*
M9	1717	−1829	Oph	7	Gl. Cl.	
M10	1656	−0404	Oph	7	Gl. Cl.	
M11	1849	−0618	Sct	6	Op. Cl.	
M12	1646	−0154	Oph	7	Gl. Cl.	
M13	1641	+3630	Her	6	Gl. Cl.	*
M14	1736	−0314	Oph	8	Gl. Cl.	*
M15	2132	+1202	Peg	6	Gl. Cl.	
M16	1817	−1347	Ser	6	Op. Cl.	*
M17	1818	−1611	Sgr	8	D. Neb.	*
M18	1818	−1708	Sgr	8	Op. Cl.	
M19	1701	−2613	Oph	7	Gl. Cl.	
M20	1800	−2302	Sgr	6	D. Neb.	*
M21	1803	−2230	Sgr	7	Op. Cl.	
M22	1834	−2357	Sgr	6	Gl. Cl.	
M23	1755	−1901	Sgr	7	Op. Cl.	
M24	1817	−1826	Sgr	5	Op. Cl.	
M25	1830	−1916	Sgr		Op. Cl.	
M26	1844	−0926	Sct	9	Op. Cl.	
M27	1958	+2238	Vul	8	P. Neb.	*
M28	1823	−2453	Sgr	7	Gl. Cl.	
M29	2023	+3825	Cyg	7	Op. Cl.	
M30	2139	−2320	Cap	8	Gl. Cl.	
M31	0041	+4107	And	5	Sp. Gx.	*
M32	0041	+4043	And	9	El. Gx.	
M33	0132	+3030	Tri	7	Sp. Gx.	
M34	0240	+4239	Per	6	Op. Cl.	
M35	0607	+2420	Gem	5	Op. Cl.	
M36	0533	+3408	Aur	6	Op. Cl.	
M37	0550	+3233	Aur	6	Op. Cl.	
M38	0527	+3549	Aur	7	Op. Cl.	
M39	2132	+4818	Cyg	6	Op. Cl.	
M40	−	−	−	−	−	
M41	0646	−2044	CMa	5	Op. Cl.	
M42	0534	−0524	Orion	6	D. Neb.	*
M43	0534	−0517	Orion	9	D. Neb.	
M44	0838	+1948	Cnc	4	Op. Cl.	*
M45	0345	+2402	Tau	2	Op. Cl.	*
M46	0741	−1445	Pup	6	Op. Cl.	
M47	−	−	−	−	−	
M48	0812	−0148	Hya	−	Op. Cl.	
M49	1228	+0809	Vir	9	El. Gx.	
M50	0702	−0818	Mon	6	Op. Cl.	
M51	1329	+4721	CVn	8	Sp. Gx.	*
M52	2323	+6126	Cas	7	Op. Cl.	
M53	1312	+1820	Com	8	Gl. Cl.	
M54	1853	−3031	Sgr	8	Gl. Cl.	
M55	1938	−3100	Sgr	5	Gl. Cl.	
M56	1916	+3007	Lyr	8	Gl. Cl.	
M57	1853	+3300	Lyr	9	P. Neb.	*
M58	1235	+1158	Vir	9	Sp. Gx.	
M59	1241	+1148	Vir	10	El. Gx.	
M60	1242	+1143	Vir	9	El. Gx.	
M61	1220	+0438	Vir	10	Sp. Gx.	
M62	1659	−3005	Oph	7	Gl. Cl.	
M63	1315	+4211	CVn	10	Sp. Gx.	
M64	1255	+2141	Com	9	Sp. Gx.	*
M65	1117	+1317	Leo	9	Sp. Gx.	
M66	1119	+1310	Leo	8	Sp. Gx.	*
M67	0849	+1155	Cnc	6	Op. Cl.	
M68	1238	−2636	Hya	8	Gl. Cl.	
M69	1829	−3222	Sgr	9	Gl. Cl.	
M70	1841	−3220	Sgr	10	Gl. Cl.	
M71	1952	+1836	Sga	9	Gl. Cl.	
M72	2052	−1239	Aqr	10	Gl. Cl.	
M73	−	−	−	−	−	
M74	0135	+1538	Psc	10	Sp. Gx.	
M75	2004	−2201	Sgr	8	Gl. Cl.	
M76	0140	+5126	Per	12	P. Neb.	
M77	0241	−0009	Cet	9	Sp. Gx.	
M78	0545	+0003	Ori	10	D. Neb.	
M79	0523	−2433	Lip	8	Gl. Cl.	
M80	1615	−2255	Sco	8	Gl. Cl.	
M81	0954	+6912	UMa	8	Sp. Gx.	
M82	0954	+6950	UMa	9	Sp. Gx.	
M83	1335	−2943	Hya	10	Sp. Gx.	
M84	1224	+1303	Vir	9	El. Gx.	
M85	1224	+1821	Com	9	El. Gx.	
M86	1225	+1306	Vir	10	El. Gx.	
M87	1229	+1233	Vir	9	El. Gx.	
M88	1231	+1435	Com	10	Sp. Gx.	
M89	1234	+1243	Vir	10	El. Gx.	
M90	1234	+1319	Vir	10	Sp. Gx.	
M91	−	−	−	−	−	
M92	1717	+4311	Her	6	Gl. Cl.	
M93	0742	−2348	Pup	6	Op. Cl.	
M94	1250	+4117	CVn	8	Sp. Gx.	
M95	1042	+1152	Leo	10	Sp. Gx.	
M96	1045	+1159	Leo	9	Sp. Gx.	
M97	1113	+5512	UMa	12	P. Neb.	
M98	1212	+1504	Com	11	Sp. Gx.	
M99	1217	+1435	Com	10	Sp. Gx.	
M100	1221	+1559	Com	11	Sp. Gx.	
M101	1402	+5429	UMa	10	Sp. Gx.	
M102	1506	+5557	Dra	11	Sp. Gx.	
M103	0131	+6033	Cas	7	Op. Cl.	
M104	1238	−1128	Vir	9	Sp. Gx.	*
M105	1046	+1245	Leo	9	Sp. Gx.	
M106	1218	+4728	CVn	7	Sp. Gx.	
M107	1631	−1259	Oph	9	Gl. Cl.	
M108	1110	+5551	UMa	10	Sp. Gx.	
M109	1156	+5332	UMa	11	Sp. Gx.	

*Denotes well-known objects of special interest.

image) screen. This improves apparent image brightness and makes focusing much easier and more precise.

Use an air-release cable to trip the camera's shutter. Manually retract the camera's instant-return mirror and wait a few seconds for the vibration to damp out prior to making the exposure.

Exposure times will vary greatly depending upon subject brightness, the effective focal length and resulting f-number, and film sensitivity (ASA rating). When using the Tele-Extender with an 18-mm ocular and ASA 200 film, the exposure time for Jupiter and the lunar terminator is approximately 1 to 2 seconds; for Saturn, about 4 seconds; for Mars at a close opposition, about ½ second; and for Venus, about 1/15 second.

DEEP SKY PHOTOGRAPHY

Start out with lunar and planetary photography before taking on the art of deep sky photography. Do not become discouraged, as this is really a trial and error learning experience.

The brightness of stellar and nebulous images at the focal plane is not governed by the same rule. The brightness of a star is determined by the square of your telescope's aperture. A star is four times brighter in a 2-inch telescope than it is in a 1-inch telescope. A nebula is not necessarily four times brighter.

The brightness of nebulae depends on the square of the focal ratio or f-number of your telescope. Nebulae (and many star clusters) appear in your telescope as virtually uninterrupted areas of light—not point sources. The larger your f-number, the dimmer the images of these objects are. A nebula is four times brighter at f/5 than it is at f/10. The brightness of celestial images as they appear on film depends on film speed or ASA rating. A film rated ASA 400 is four times faster, or more sensitive to light, than a film rated ASA 100. If you see a picture of a nebula made at f/5 with a 10-minute

exposure on ASA 100 film, can you get the same image density at f/10 in a 10-minute exposure on ASA 400 film? You will probably get a similar density if you use a film of the same "color" and if the atmospheric conditions are equivalent. Your film, however, will be faster and grainier, so you will lose some detail.

What about making a 40-minute exposure at f/10 with the same type of ASA 100 film used in the original photo? You will probably get less image density because of a reciprocity failure. This is the inability of film to respond as well to low levels of light over longer periods as it does to higher levels of light over shorter periods.

Kodak makes special 103a series spectroscopic films (i.e., low reciprocity failure) specifically for astrophotography. These films—103aE (red sensitive), 103aO (blue sensitive), and 103aF (panchromatic)—are available from several companies that advertise in astronomical magazines. Type 103aO is best for galaxies and reflection nebulae, 103aE is best for diffuse (emission) and planetary nebulae, and 103aF is good for all deep sky objects.

USING THE OFF-AXIS GUIDING SYSTEM

While lunar and planetary photographs are essentially snapshots, time exposures are required to photograph star clusters, nebulae, and galaxies. The times will vary depending upon the telescope being used. The deep sky photographer must contend with *image drift* that is of little concern to the visual observer or to the planetary photographer who has lined up with the celestial pole. Image drift is caused by misalignment with the pole, atmospheric scintillation, and by long-period irregularities inherent in the mechanics of any clock drive.

To guide your telescope through a time exposure, you need a way to establish an in-the-field reference for image drift. You also need a way to guide at a much higher power than is equivalent to the image scale at which you are shooting. All methods incorporate some form of off-axis guiding system. The accessory for the Celestron system incorporates a tiny prism to divert light from a star at the edge of your photographic field up into a high-power 12.5-mm ocular with illuminated cross hairs.

After attaching your camera, locating the object, and focusing, look through the finder telescope for a suitable guide star near the subject. Loosen the slip ring of the guider. Rotate the guider until the eyepiece neck is pointing in the same direction as the guide star appears to be through the finderscope. Only a small amount of searching will be necessary to get the guide star centered in the field of the guiding eyepiece. The right ascension and declination slow motions may be used for this purpose. If you can't see a potential guide star through the finderscope, it will be too faint to use successfully.

Focus on the guide star by raising or lowering the guiding ocular. You should focus the image of your subject on the ground glass of the camera before focusing the guiding ocular. Using Celestron's optional drive corrector, keep the guide star centered on the intersection of the cross hairs for the duration of the exposure. Many people find it helpful to orient the cross hairs so that movement along one cross hair is in right ascension and movement along the other is in declination.

Chapter 7

A Colorful Universe

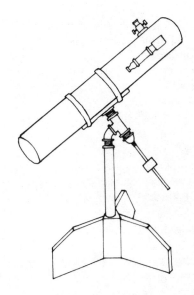

To MOST PEOPLE WHO ARE NOT FAMILIAR WITH deep sky phenomena, the night sky is merely black and white—the white stars against the black sky. There may be a hint or two of blue light from some stars, but there are few colors to behold.

Viewing the heavens through a telescope will immediately bring out some coloration in the seemingly black and white skies—the reddish tint of Mars, the myriad of colors in Saturn and Jupiter, and more. Many colors present in star clusters, nebulae, and spiral galaxies are much dimmer than the white light you so readily perceive. A picture must be taken using sophisticated astrophotographic equipment.

Most color films do not respond evenly to different colors. This is not terribly important for terrestrial photography. For deep space applications, though, response to many light levels and colors is paramount.

The beautiful photographs of near and deep space on pages 105-120 were supplied by Celestron International, Torrance, California, U.S.A. These pictures were shot under ideal viewing conditions.

A dry ice camera was probably used to produce most of these photographs. This is a special camera that is packed with dry ice to freeze the film to be exposed. When cooled to very low temperatures, color film is much more sensitive to very faint traces of incoming light. Equal photographic results will not be obtained in your backyard on a hot summer's evening using the same telescope and your 35-mm camera. Don't be too disappointed, though. With a little study, some additional photographic equipment, and accessories, you may have excellent results if you choose your location correctly.

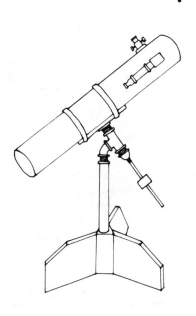

Chapter 8

Building Your Own Reflector Telescope

MANY AMATEUR ASTRONOMERS ELECT TO BUILD their own telescope instead of purchasing manufactured and aligned instruments. You can build a reflector telescope as good as many commercial types. Buy finished materials and save the labor costs by doing the assembly yourself. The most critical parts of a reflecting telescope include the primary and secondary mirrors. These can be purchased as finished components for mounting, or you may elect to grind your own primary mirror (see Chapter 9). Most companies that sell the mirror blanks for grinding provide complete instructions. A few astronomers have elected to cast their own mirrors. This is an even more complex procedure that can take months.

The reflector telescope design is the easiest to build yourself. Refractor telescopes require the grinding of several lens surfaces. Buying kits for refractor telescopes is most difficult. The reflector

telescope is a simple design, and its performance will rival that of the refractor. A good 4-inch refractor telescope can cost more than $600 when purchased commercially, while a 4-inch reflector telescope can cost less than $300.

PARTS FOR A REFLECTOR TELESCOPE

Figure 8-1 is a diagram of a typical reflector telescope. The objective mirror is the heart of this instrument. If the primary mirror is not right, the entire package will not work properly. Most beginning telescope builders will choose a 4¼- or 6-inch (mirror diameter) for their first design, although a few may elect to build 8-, 10-, or even 12-inch telescopes. When the objective mirror size exceeds 6 inches, the size and complexity of the telescope mount becomes much greater.

The diagonal mirror directs the image from the objective mirror into the eyepiece. This component

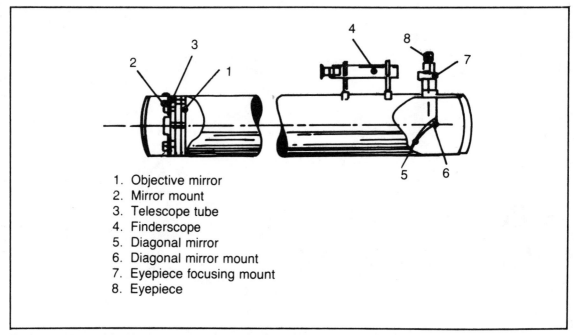

1. Objective mirror
2. Mirror mount
3. Telescope tube
4. Finderscope
5. Diagonal mirror
6. Diagonal mirror mount
7. Eyepiece focusing mount
8. Eyepiece

Fig. 8-1. A reflector telescope utilizes mirrors rather than lenses.

must be optically flat and should be aluminized and overcoated. The best diagonal mirrors are polished to an accuracy of at least ⅛ wave. They are quite difficult to make because of their small size, and they are almost always purchased in finished form.

The mirror mount can be slightly larger than the primary mirror. Mirror mounts must be rigid while being able to adjust the mirror's angle for the alignment procedures. These can be made from wood or metal or purchased outright. The diagonal mirror mount can also be constructed, but the diagonal mirror can be purchased already mounted in place.

The telescope tube is a means of mechanical support and is not essential to telescope operation. Many large telescopes do away with the enclosing tube completely and align the primary and secondary mirrors by means of complex, open mechanical structures that are mounted to the same base. This

design should not be chosen by the beginning telescope maker because of the complexities involved. The main purpose for eliminating the tube is to reduce the instrument's weight. Special mounting arrangements are necessary for the open design, and stray light entering broadside to the telescope can create viewing difficulties.

The telescope tube can be made of cardboard, wood, fiberglass, aluminum, or other materials. It should be painted with a matte black finish on the inside to reduce light reflections to a minimum. The inside diameter of the tube should be about 15 to 20 percent larger than the diameter of the primary mirror.

The eyepiece focusing mount is important for ease of viewing. Some telescopes are made with a friction sleeve that allows the lens to be inserted and moved up and down by hand for focusing. This makes fine focusing difficult. Fortunately, eyepiece mounts for reflecting telescopes are available through many astronomical outlets.

The eyepiece is usually purchased from an astronomical equipment company. Choose an eyepiece mount that will accept an eyepiece of the chosen diameter.

A tripod can be easily constructed from wood, aluminum, and angle iron, although it is most difficult to construct one with motor drive. Motor drive for compensation of the earth's rotation is a highly touted feature of any telescope, but it is unnecessary for many types of viewing. If you plan to use your instrument for casual viewing, especially short observations, you can get by without the motor drive. You must readjust the direction of the telescope every few minutes as the star or planet being observed begins to drift from the aperture. If you plan to use your instrument for long and detailed studies of the heavens or for astrophotography, then a tracking drive is mandatory. You can usually purchase a commercial mount for about the same price as the materials needed to construct one.

A REFLECTOR TELESCOPE USING PREFABRICATED COMPONENTS

Components from Edmund Scientific Company are available individually or in a kit. Edmund also includes a handy chart that allows you to select the exact components needed for a reflector telescope of a specific aperture.

The first consideration is the primary and diagonal mirrors. If you want to build a 6-inch, f/8 telescope, you would choose an appropriate parabolic mirror. Edmund Scientific lists a completed Pyrex brand heat-resistant glass mirror of a 48-inch focal length for $79.95. Select an appropriate diagonal mirror, which sells for approximately $29.50. Mounts must be ordered for both mirrors at a cost of $54.95 each.

The phenolic plastic tube for the telescope

costs about $50. Alternately, a spiral paper tube may be ordered that sells for $13. Cut this tube to 48 inches. An eyepiece holder costs about $20.

While it is a simple matter to build a relatively stable tripod from angle iron or even wood, the building of a stable clock drive mechanism is beyond the abilities of many amateur astronomers. Later in this chapter we will discuss replacing many prefabricated components with home-built ones. When a clock drive is necessary, we recommend that it be purchased from a reputable manufacturer.

Assembly is quite simple. Start by cutting the telescope tube to size. The desired length for an f/8 instrument is 48 inches.

Using the primary and diagonal mounts supplied by a manufacturer, it is simple to mount each in opposite ends of the hollow tube. These mounts may be friction-fitted into each end. Others will require four holes to be drilled equidistantly at each end of the tube. The circular mounts are then held in place by setscrews after the appropriate mirrors have been attached. If your particular tube has not been precut to accept the appropriate eyepiece holder, this should be done before the mirror mounts are installed. This keeps the mirror surfaces free of shredded debris and also avoids any possibility of damaging the mirror surfaces.

The hole that is cut to fit the eyepiece mount used with this project (Edmund Scientific number 50,077) should be circular with a diameter of 1½ inches. Figure 8-2 shows the template for installing this particular mount. The irregularity in the circular hole is required when using a different diagonal mirror mount that consists of a shaft extending from the eyepiece mount downward into the scope tube, at the end of which the secondary mirror is mounted. If you use the mount specified, it will not be necessary to include this irregularity. Four holes are drilled equidistantly around the circular hole to allow the eyepiece mount to be attached to the tube.

When the eyepiece mount has been affixed to

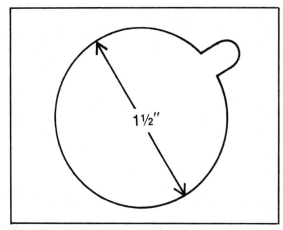

Fig. 8-2. A 1½-inch diameter hole is cut using a template.

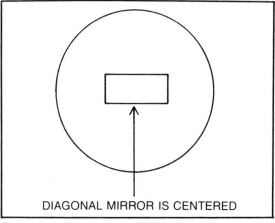

DIAGONAL MIRROR IS CENTERED

Fig. 8-3. Proper alignment is indicated when this view appears.

the tube, insert the diagonal mirror mount and align the diagonal with the eyepiece aperture. With the eyepiece removed, proper alignment is indicated when the view appears as shown in Fig. 8-3. When this has been accomplished, the basic telescope assembly is complete. Insert the eyepiece in the holder and attach the tube assembly to an appropriate mount.

You may experience minor problems with eyepiece alignment and the drilling of the telescope tube. The latter is not all that critical. If you make a mistake, start over again on the other side of the tube. The bogus hole can then be covered with masking tape without creating any viewing difficulties.

TELESCOPE TUBE

Perhaps you want to build almost all components. The most expensive component of a reflecting telescope can be the primary mirror mount. This can be constructed in a few hours out of readily available materials. The diagonal mirror mount will be equally expensive for many models. This can be constructed at home for less than $1 if you have a few materials on hand.

A telescope tube should be waterproof and fairly rigid. Some of the lightweight plastic pipe available at most plumbing outlets may be ideal for telescope building. These are usually found in diameters of 2, 4, 6, 8, and 10 inches. The diameter is really unimportant if it will accommodate the primary mirror. For a 6-inch mirror design, the 8-inch tubing will probably be required, although you should check the fit in a 6-inch diameter length. Perhaps the mirror mount (which will also be home-built) can be made to accommodate the smaller size. This pipe will cost $6 to $8 per foot for the larger sizes. The minimum length is usually 8 to 10 feet.

Your plumbing retailer may not sell you less than a full section. Most large plumbing outlets often have odd sections of this material that have been returned due to defects or other reasons. You may be able to pick up one of these odd lengths for a few dollars. Your local plumbing outlet may be able to cut it to the proper length for you and drill the required holes.

The weight of certain types of plastic pipe may mean a larger mounting tripod, so consider this when shopping around. Don't make the mistake of purchasing the ultralightweight piping, as it may be too flexible for your design.

Paper tubing is available. Printing outlets may have a supply of circular cardboard tubing around which large quantities of paper were originally wrapped.

Make sure that this material is not too flexible for astronomy uses. Paint the outside and inside of the tube with a matte black finish. Outdoor paint or varnish is suitable. This will prevent unwanted reflection inside the tube, and the outside coat will partially protect the assembly from moisture.

If you know something about fiberglassing techniques, you can construct your own lightweight fiberglass tube using a circular mold. People at a local fiberglass fabrication outlet in your town may be able to make you one at a reasonable price. Lightweight fiberglass makes an ideal telescope tube assembly and can be custom-manufactured to include the cutouts you desire for your project.

There is no rule that says the housing assembly must be a circular tube. Some telescopes use no tube at all and simply mount their mirrors on an open frame. A rectangular-shaped enclosure that is approximately 48 inches long and allows for the mounting of the 6-inch mirror is completely acceptable. It may be easier to attach to a home-built mount due to the flat surface provided. This can be easily constructed from four plywood sections that are fitted together to form an enclosed housing and then weatherproofed. This may not look as pleasing as the circular tube, but it's performance that counts.

You may want an open tube design (Fig. 8-4). Two 6½-inch or larger metal rings are attached to the ends of four aluminum tubing sections. The ends of the tube sections can be threaded for attachment to the rings, or they may be drilled to allow a small bolt to serve as the attachment point. A local metalworking shop can probably supply the rings, and a hardware store can probably provide the lightweight aluminum tubing. Aluminum is ideal because of its light weight. Other metallic materials such as brass may be used. When aluminum is used, the low weight of the finished assembly will probably surprise you. The open tube design requires more active maintenance and cleaning of the mirror surfaces, as they are exposed more to dust. This type of telescope should be stored in an enclosed case when not in use.

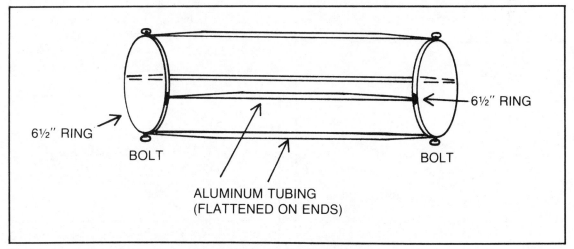

Fig. 8-4. An open tube design may be chosen. Aluminum tubing is used here.

PRIMARY MIRROR MOUNT

The primary mirror mount can cost more than $300 for larger mirrors. You can make a suitable one at home that costs only a few dollars. You need two small sections of ¾-inch plywood, a few small angle brackets made from brass or steel, and a jigsaw. Figure 8-5 shows the measurements of the plywood surface that mate with the primary mirror. The measurements are for the mounting of a 6-inch mirror. A circular piece of plywood slightly larger than the 6-inch mirror is cut. The diameter of the circular plywood slab is 6⅛ inches. To give yourself an outline to cut from, use a larger compass calibrated to a radius of about 3 1/16 inch. Draw a perfect circle. Cut the circle out with a jigsaw. Cut a matching section from some felt material. This section is placed over the circular plywood section and bonded in place with glue. This felt pad acts as a shock absorber, preventing scratching of the bottom mirror section and eliminating vibrational effects upon viewing.

Drill three holes as shown in Fig. 8-6. The holes should be wide enough to accommodate a ¼-inch diameter bolt. Chip out a recess area on the mirror side of the circular slab to allow for the bolt heads to fit beneath the mounting surface. These recesses should be deep enough to prevent the bolt

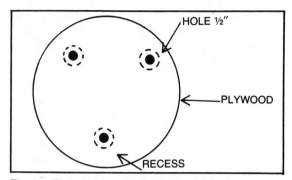

Fig. 8-6. Three holes are drilled in the plywood.

head from coming in contact with the mirror when tightly fitted against the felt surface.

Cut another circular slab of plywood that is the same width as the inside diameter of your telescope tube. You should cut this portion a bit oversized than undersized. An oversized section can be reduced in diameter with coarse sandpaper if necessary. This section should be slightly larger than the first. Lay the first section in the middle of the second. Mark the location of the first section's three holes on the lower piece. Insert a small nail through the holes of the top section, thus making an indentation in the second one.

Drill holes in the second section as you did with the first, but do not include the recesses. Returning to the original felt-covered section, drill three 1/16-inch holes to a depth of about 1/8 inch equidistant along the outer rim (Fig. 8-7). Prepare

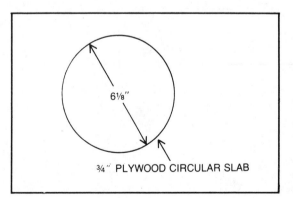

Fig. 8-5. The measurements of this plywood surface mate with the primary mirror.

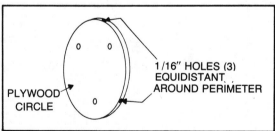

Fig. 8-7. Drill three 1/16-inch holes equidistant along the outer rim.

three small brackets—each 2 inches in length—from small gauge angle stock. Small holes should be drilled in one end of each clip. Form a 90-degree bend at ½ inch at the other end of each clip. The finished product should look similar to Fig. 8-8. From the original felt material, cut three small pads and cement them to the angled portion of the bracket. Attach each bracket with a small wood screw to the circular slab. Do not tighten the screws

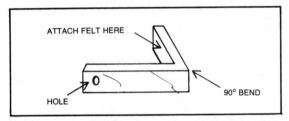

Fig. 8-8. Form a 90-degree bend at the end of each bracket.

all the way. Some flexibility must be allowed for inserting the mirror.

Insert three 2½-3 inch carriage bolts in the three previously drilled holes. Each bolt is fitted with a small ½-inch diameter spring about 1 inch long. The entire assembly is then mated to the second section of circular plywood (Fig. 8-9). On the other side of the second section, cover the threaded tips of each carriage bolt with a small washer. Attach a mating wing nut to each bolt.

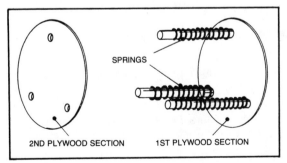

Fig. 8-9. The entire assembly is mated to the other plywood section.

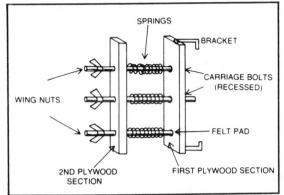

Fig. 8-10. The two sections are tightened by bolts.

Tighten the wing nut until the first section of plywood is held firmly in place, but not against the second section.

Carefully mount the mirror onto the bracketed section of plywood. If you tightened the bracket screws all the way, loosen them a bit to allow the brackets to be slightly separated. Slip the mirror between the brackets until it rests on the felt surface. Take care not to scratch the mirror's reflecting surface. Tighten the brackets until they hold the mirror firmly (Fig. 8-11). If the brackets do not fit tightly to the mirror surface at their 90-degree bends, drill mounting holes in the brackets' lower sections. Do this only after the mirror has been removed and is safely out of the way. The brackets should hold the mirror firmly in place, but not so tightly that they present damaging stresses to the mirror's surface.

When a proper fit has been obtained, loosen the bracket screws one more time and remove the mirror. Drill three small equally spaced holes in the outer rim of the second or bottom plywood section. These holes are used to mount the entire assembly at the bottom of the telescope tube. Drill the tube with three equidistant holes first, insert the assembly, and make indentations in the outer edge of the plywood section for drilling guides.

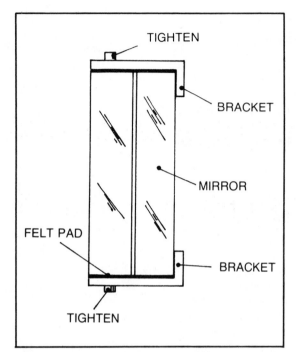

Fig. 8-11. The mirror is bolted firmly in place.

Remount the mirror as described earlier and insert the entire assembly into the telescope tube. Rotate the assembly until the guide holes in the endmost plywood section align with the holes in the telescope tube. Permanently attach the primary mirror mount using number 8¾-inch screws. Your mirror mount project is now complete. The wing nuts will be used to vary the angle of the primary mirror during the alignment process.

Installation of the mirror and completed mirror mount should come last. The eyepiece holder and diagonal mount should be installed before the primary mirror is mounted. If these steps are reversed, a slip of a component at the diagonal mirror end of the tube could cause the primary mirror to become damaged.

Because wood has a tendency to swell when it collects moisture, you will probably have to realign the primary mirror periodically. This is a relatively simple process using the wing nut adjustment. Persons living in extremely humid climates will probably want to apply a moisture-proofing compound to the wooden components.

The assembly is fairly straightforward and is not that difficult. It is nearly impossible to ruin the plywood sections through drilling errors. If you make a mistake in drilling, simply drill another hole. If the mistake occurs on one of the edges, just rotate the slab and try again. The main concern is mechanical rigidity. If the brackets do not seem to be holding the mirror to your satisfaction, make the proper adjustments.

You can simply replace the ¾-inch plywood sections with ⅛-inch aluminum sheets or brass stock. If weight becomes a factor, you can punch numerous holes in the sections. Special bolts can be recessed into the aluminum material, or simply cover the bolt heads with plenty of felt. Make certain the brackets do not press the mirror too firmly against the bolt heads.

This project may give you other ideas of how to build your own primary mirror mount. Be aware of the stringent requirements for mechanical rigidity. This applies doubly when larger mirrors are to be utilized in telescope construction. An 8-inch mirror will probably require about 50 percent more holding rigidity from its mounting assembly. The stresses increase greatly past the 8-inch point. If you construct a reflector with a 10 or 12-inch mirror, give the mounting the attention it deserves.

EYEPIECE AND DIAGONAL MOUNT

The eyepiece holder is considered by many to be the most difficult part of telescope construction. This may be true when it is necessary to have rack and pinion focusing, but simple slide-focused eyepiece holders can be made very easily and quickly. If you want rack and pinion focusing, purchase a manufactured unit.

Do-it-yourself builders usually include the diagonal mirror mount as a part of the eyepiece

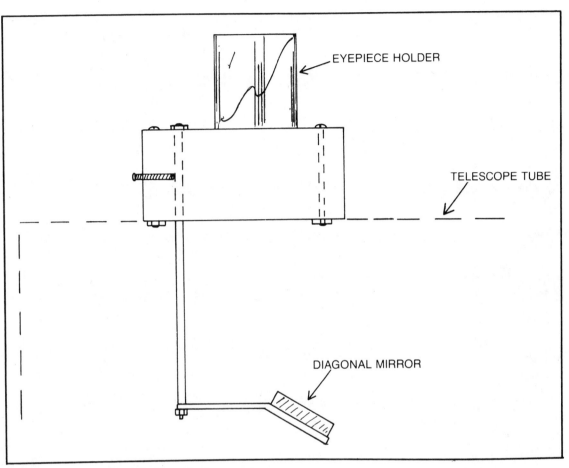

EYEPIECE HOLDER

TELESCOPE TUBE

DIAGONAL MIRROR

Fig. 8-12. Eyepiece holder with diagonal mirror mount.

holder. While spider-type diagonal mounts provide better rigidity, the mirror is small, lightweight, and lends itself readily to less extravagant mounting configurations. Figure 8-12 shows a one-piece eyepiece holder and diagonal mirror mount of wood and metal construction. The diagonal mirror mount consists of a 4½-inch steel post that is attached directly to the eyepiece holder mounting block. The other end is fitted through a rectangular angle bracket of the same diameter as the diagonal mirror and bent at a 45-degree angle. The depth of the

diagonal in the telescope tube is adjusted by a screw mounted to the eyepiece holder block. This can be considered as a one-piece assembly, although the diagonal mounting can be removed for cleaning by simply taking out the setscrew.

Start by fabricating the mounting block. This is made from a 2×3×1-inch section of block (Fig. 8-13). Cut a circular hole 1 5/16 inch in diameter in the center of this block. This will be slightly small for the insertion of the focusing tube, but it can be enlarged during a later step by sanding. Five holes

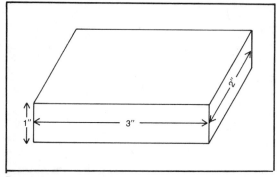

Fig. 8-13. The mounting block is constructed from a 2×3×1-inch section of wood.

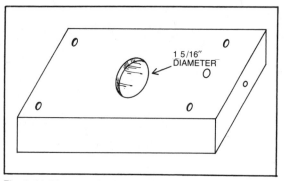

Fig. 8-14. The completed block.

are drilled in the block to allow for mounting to the telescope tube and for inserting the mounting rod for the diagonal mirror. Four holes (one in each corner of the block) allow for the insertion of screws or bolts for the block attachment, while the fifth hole (drilled ¼ inch from the edge of the center hole) will later be fitted with the mounting post. Another hole that is drilled in the end of the block will connect with the mounting post hole. Figure 8-14 shows the completed block with all holes drilled.

Obtain a steel rod approximately 3/16 inch in diameter. The rod should be at least 4¼ inches long

to allow for lateral adjustment. If you can't obtain a rod with a threaded end from a hardware store, cut threads on one end for attachment of the angle bracket to which the diagonal mirror is mounted. The angle bracket should be bent at a 90-degree angle ¾ inch from one end. This ¾-inch section is drilled to accept the steel post. Figure 8-15 shows how the finished product should look with the post inserted and attached with a small nut.

Cut a 3-inch section of 1 5/16-inch diameter aluminum tubing. Make a cut in the pipe 2 inches long at one end (Fig. 8-16). Use a hacksaw placed longitudinally along the tubing. At the bottom of this

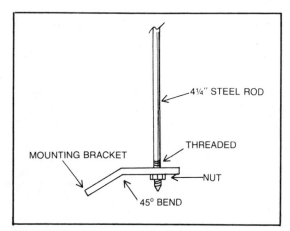

Fig. 8-15. The finished product with the post inserted.

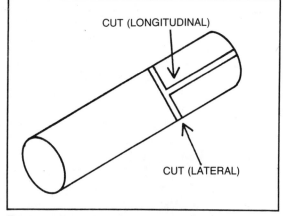

Fig. 8-16. Make a cut in the aluminum tubing 2 inches long.

cut, make another cut laterally. Insert the uncut end of the tubing into the hole provided for it in the wooden block. This will probably take some sanding and gouging, but it should not take long to get this end mounted flush with the bottom of the block. Take care not to bend the tubing, especially the portion that has been cut. If you have the central hole too wide, simply wrap the last 1 inch of the tubing with a layer of electrical tape. Some cement on the inserted portion of the tubing will assure bond with the wood. If the center hole was drilled properly, though, a tight friction fit will be obtained. No further bonding will be necessary.

This eyepiece holder is designed for standard 1¼-inch eyepieces. Carefully insert it into the remaining end of the tube after all burrs from the cutting process have been removed. It should be a tight fit, with the slotted section giving just enough to allow for insertion. If not, press firmly on both sides of the cut to reduce the diameter or insert a flathead screwdriver in the cut and twist to slightly widen it. Alternately, you can fit the top two portions with an adjustable clamp for a tighter fit, although lens adjustment will involve more steps.

The eyepiece holder at this point should appear similar to Fig. 8-17. To mount the holder to the telescope tube, cut a hole as shown in Fig. 8-18. The irregularity in this otherwise circular cut is for the insertion of the diagonal mounting post. Place the wooden block squarely over the opening (a flashlight beam directed into the tube opening can help with alignment). Insert a small nail into each previously drilled hole to mark the scope tube for drilling. Drill these holes from the guides provided.

The actual means of attachment to the tube will depend upon the type of material used. Wood or metal screws may be appropriate, or you can use a standard nut and bolt arrangement. Attach the block and holder firmly to the telescope tube.

Insert the diagonal mirror mount rod through the cutout provided and into the allotted hole in the wood block. Insert a small screw into the end of the

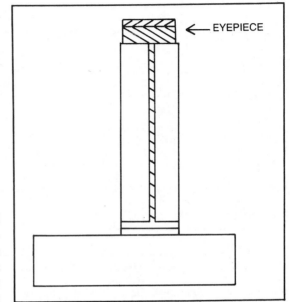

Fig. 8-17. View of the eyepiece holder.

block until it presses firmly against this rod. You may have to make minor adjustments to get a good fit. No movement should be detected when the scope tube assembly is vigorously shaken.

When the proper mating has been accomplished, slightly loosen the screw and remove the rod and bracket. You can now mount the diagonal mirror. The mirror will require some adjustment. A light coating of putty or other adhesive material can be used for the test stages. Apply the putty to the surface of the angle bracket. Mount the mirror as would normally be done in a permanent installation. Carefully reinsert the diagonal post through the tube and into the wood block, binding it with the setscrew. The center of the mirror should be about 3½ inches from the side of the telescope tube containing the eyepiece holder. Peer through the holder with the eyepiece removed. You should see something similar to Fig. 8-19. If not, you may have to alter the angle of the bend in the bracket or slightly bend the diagonal post. This will involve

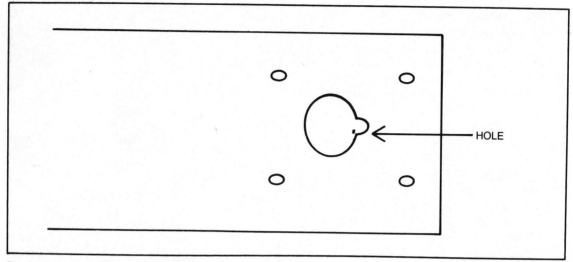

Fig. 8-18. A hole is cut to mount the holder to the telescope tube.

removing the mounting from the scope tube, removing the diagonal mirror, and then making the adjustments. Be very careful in removing this assembly, because the mirror is only held in place by the temporary adhesive.

You may then reinsert the diagonal mirror and its mounting and install the primary mirror mounting. This should be done with the telescope tube pointed downward. If the diagonal mirror should drop from its mounting, it will fall downward toward the ground and not toward the expensive primary mirror. A soft surface should be laid beneath the tube to protect the diagonal mirror in case it slips out. When the two mirrors have been installed, the instrument may be collimated as described in Chapter 11. This is done before the diagonal mirror is permanently mounted in case any additional bending of the bracket angle is required. When collimation has been accomplished, remove the diagonal mirror assembly from the tube and clean all adhesive residue from both the mirror and the bracket. The diagonal is now permanently attached to the bracket by using a commercial bonding cement. This reattachment may induce some vertical or horizontal differences in the diagonal, but these should be negligible and can be corrected by adjusting the mounting post.

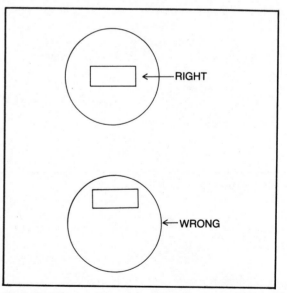

Fig. 8-19. The view through the holder should appear as in the top drawing. An incorrect view is shown in the bottom drawing.

MOUNTING PLATFORMS

For the 6-inch telescope, many medium-duty mounts are available that offer excellent mechanical stability, along with the mounting brackets and the clock drive you will probably need. Edmund Scientific advertises a medium-duty mount with clock drive and pedestal for less than $175. It is most difficult for home astronomers to build good mounts. This usually requires the machining of special parts, strict adherence to angles, and many other criteria. If a clock drive is to be installed, you must buy a 1/15-rpm motor and matching gear drives.

A Celestron C90 telescope is a very compact and efficient instrument measuring about 1 foot in length. It can easily be held in one hand. This applies to only the scope tube and its mirrors (Fig. 8-20). When this assembly is attached to the declination setting circles, the fork tine, and the drive base, the overall package gets cumbersome. When the assembly is attached to the equatorial wedge and tripod, it becomes even more cumbersome.

While all this equipment is mandatory for long observations and astrophotography or for finding a particular object location, much of this apparatus can be eliminated for casual viewing. Casual viewing probably encompasses more than 90 percent of amateur astronomy observing.

Using the camera tripod with the base tele-

Fig. 8-21. A camera tripod is used with the Celestron 90.

scope is a possibility (Fig. 8-21). This requires no modifications, because the telescope is already drilled to accept the screw mounting of the camera tripod. The tripod shown in Fig. 8-21 was purchased at K-Mart. It has adjustable legs, all types of settings, and serves very well for casual viewing. Figures 8-22 through 8-26 show various configurations that can be used with this tripod for astronomical viewing. It serves quite well because of the telescope's light weight. A heavy 6-inch reflector would not work as well. For 3 and 4-inch telescopes of the reflector type, small refractors, and up to 8-inch catadioptrics, a heavy-duty camera tripod may be satisfactory.

Figure 8-27 shows a small reflector telescope with a 3-inch aperture mounted to a wooden tripod. These mounts are very popular with small aperture scopes. This one is made by Edmund Scientific.

Fig. 8-20. A Celestron 90 in its simplest form can be held easily in one hand.

Fig. 8-22. The telescope mounting plate on the tripod can be swung out 90 degrees.

Fig. 8-24. Central bar of the tripod canted 20 degrees.

Fig. 8-23. Tripod with central bar canted 45 degrees.

Fig. 8-25. The center post may be elevated for tall viewers.

Fig. 8-26. Tripod with lower legs telescoped into upper legs.

Fig. 8-27. A small reflector is mounted to a wooden tripod.

Notice the simple screw set adjustments. These could easily be fabricated in a home workshop using plumbing materials. For large telescopes, a heavier-duty tripod would be required.

For extremely large telescopes up to 10 inches, specialized mounts are made similar to the one in Fig. 8-28. This one is designed to mount the telescope to a large metal pedestal or to one made of concrete that is permanently affixed. A flange is attached to the top of the concrete pedestal for direct attachment of the mount. Figure 8-29 shows the basic structure of the permanent pedestal, while Fig. 8-30 illustrates a large telescope that has been properly mounted. Metal bands attach to the cradle and encircle the telescope at two points. This type of heavy-duty construction is required for the largest telescopes and cannot be moved from the original location.

A very simple telescope mount can be built in

your home workshop from 2-inch pipe and some wood. This design is shown in Fig. 8-31 and involves a 3-foot section of 2-inch pipe, along with a mounting cradle directly to which the telescope is attached by mounting bolts. A 90-degree elbow is used to properly position the telescope and to provide vertical and horizontal adjustments. Start by constructing the cradle, which is made from a 12×6×1-inch piece of lumber. Attach two lengths of corner molding to the top (Fig. 8-32). Drill two holes ½ inch in diameter 1½ inches from either end of the baseboard. Turn the board over and install a 2-inch floor flange in the center (Fig. 8-33). Drill the mounting holes for this flange after they are properly marked.

Rest the cradle against the center body portion of your telescope. Check that the fit is correct. These instructions will suffice for a 6-inch telescope. Mark two guides on the scope tube that align

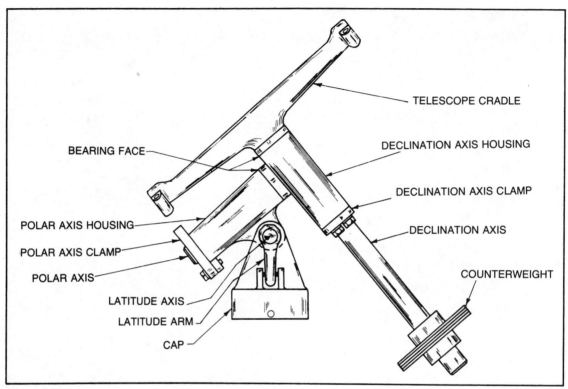

Fig. 8-28. Specialized mounts are advantageous with the larger telescopes.

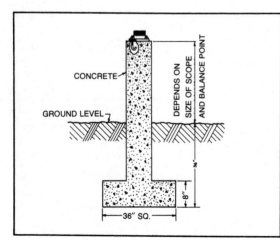

Fig. 8-29. Basic structure of the permanent pedestal.

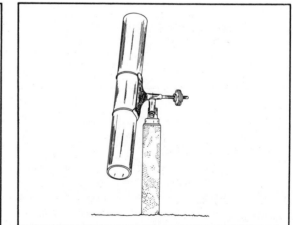

Fig. 8-30. Proper mounting configuration for a large telescope.

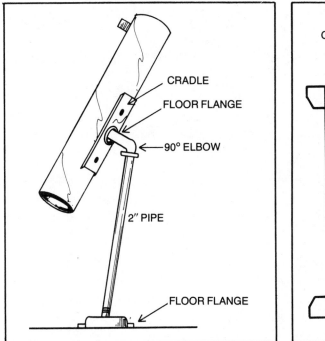

Fig. 8-31. A simple telescope mount can be constructed at home.

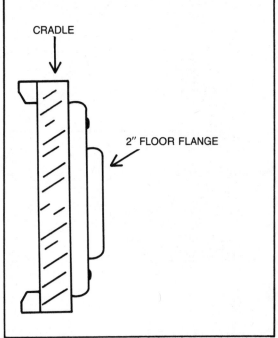

Fig. 8-32. Two lengths of corner molding are attached to the piece of wood.

with the holes in the cradle. Holes are then drilled through these guides. Two-inch bolts are inserted through the telescope from the inside out, through the cradle, and then secured by nuts (Fig. 8-34).

A 2-inch, 90-degree elbow is now screwed to the floor flange at the bottom of the cradle. Do not tighten it all the way. The other threading on the elbow attaches directly to the 3-foot length of 2-inch

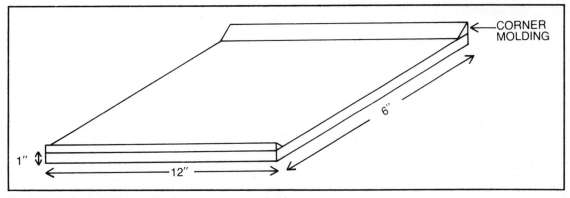

Fig. 8-31. A simple telescope mount can be constructed at home.

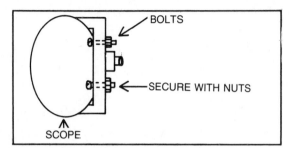

Fig. 8-34. Holes are drilled through the guides.

threaded pipe. The pipe should already be attached to the tripod base or other permanent mount.

Figure 8-35 shows how a permanent arrangement can be made by installing another floor in a concrete base measuring 2 cubic feet. Alternately,

the arrangement in Fig. 8-36 can be used. It involves attaching the floor flange to a wooden base.

This installation is recommended for permanent viewing locations only. The telescope is adjusted by rotating it vertically on the freewheeling floor flange/elbow connection at the cradle. Lateral adjustment is accomplished at the elbow pedestal connection. Enough surface-to-surface friction should be provided to prevent slipping. This is quite cumbersome and is useful only for very casual viewing.

We recommend a commercially manufactured mount and pedestal over the home-built one. The precision that has gone into the building of your telescope will be completely wasted on any but the finest of mounts.

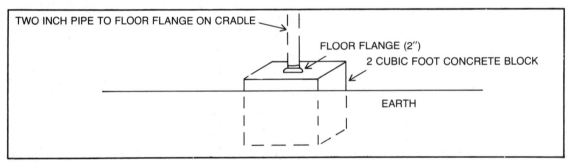

Fig. 8-35. A floor flange may be installed in a concrete base for a permanent mounting.

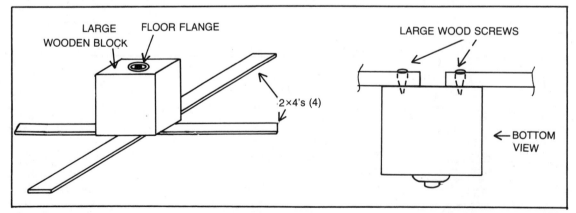

Fig. 8-36. The floor flange is attached to a wooden base.

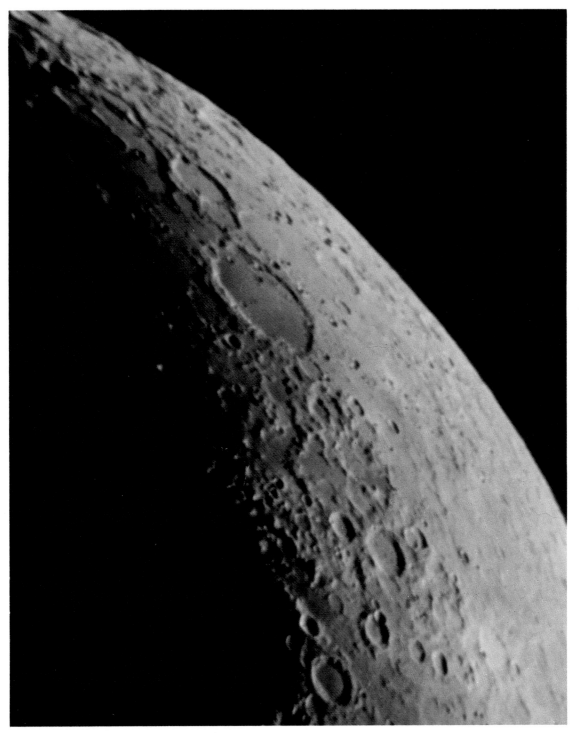

The moon as seen through a Celestron 8 telescope.

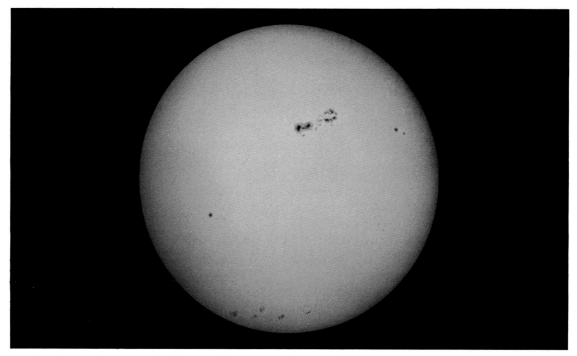

The sun and sunspots as seen through the Celestron 8.

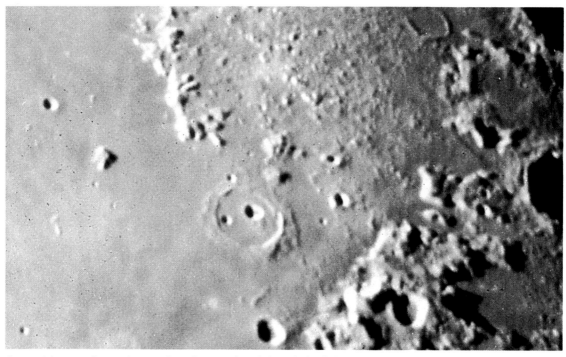

A mountainous region on the moon's surface as viewed through the Celestron 14 telescope.

The moon at half phase as seen through the Celestron 8.

The full moon at eclipse as viewed through the Celestron 8.

The Comet d'Arrest and the Helix nebula as photographed by the Celestron 8-inch Schmidt camera.

The Orion nebula is very colorful. This view is through a Celestron 8 telescope.

The Crab nebula as viewed through the Celestron 8.

The M3 globular cluster as seen through the Celestron 14.

The Andromeda galaxy is even more enchanting when seen through the Celestron 8.

The Pleiades stand out against the dimmer stars and clusters. This view is through a Celestron 8.

The Sombrero galaxy is one of the most beautiful sights in the night sky. This view is through a Celestron 8.

Globular cluster M13 as viewed through the Celestron 8.

The Andromeda galaxy as captured by the Celestron 5½-inch Schmidt camera.

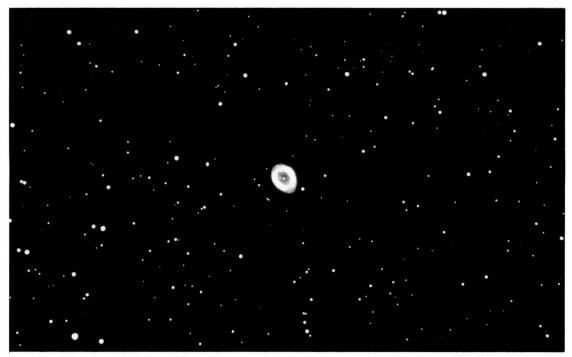

The Ring nebula as seen through the Celestron 14 telescope.

Spiral galaxy NGC 4565 as seen through the Celestron 14.

The Whirlpool galaxy as seen through the Celestron 14.

The Dumbbell nebula as viewed through the Celestron 14.

The Horsehead region of Orion as captured by the Celestron 5½-inch Schmidt camera.

Spiral galaxy M33 as seen through the Celestron 8-inch Schmidt camera.

NGC 2070 Tarantula and a large Magellanic Cloud as captured by the Celestron 8-inch Schmidt camera.

The California nebula as seen through the Celestron 8-inch Schmidt camera.

Eta Carinae as viewed through the Celestron 8-inch Schmidt camera.

118

The Lagoon and Trifid nebulae as captured by the Celestron 5½-inch Schmidt camera.

The Omega nebula as seen through the Celestron 8.

A full solar eclipse as seen through the Celestron 5.

Chapter 9

Grinding Telescope Mirrors

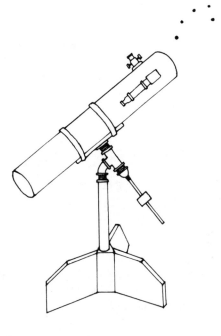

THE MOST DIFFICULT AND METICULOUS PRO-
cedure involved in building any type of telescope is grinding the glass used for the objective
and eyepiece. The reflector will require for grinding of only one piece of glass.

The component that we will be chiefly con-
cerned with in this chapter is the objective or primary mirror. This is a piece of glass that gathers
incoming light rays and reflects them to a point
where they can be magnified by the eyepiece to be
seen by your eye. The primary mirror is shaped like
a concave mirror. If the mirror is not ground to
proper measurements in a completely uniform
manner, it will not perform properly.

The theory behind forming an exact and uniform surface for the primary mirror is mathematical in nature, but it can be simply expressed. The
mirror must reflect the incoming light from an infi-
nite point so that an image is formed at a single point

inside the tube. This point is the focal point, and the
distance that this point is from the objective is the
focal length. For the mirror to perform in this manner, it must be in the shape of a parabola. This curve
is shown in Fig. 9-1.

KITS

Several companies offer complete kits for
building a reflector telescope. These kits usually
include the mirrors, both diagonal and primary, already ground in different sizes and focal lengths.
This is a viable alternative, because the builder will
only be required to do the actual assembly. Edmund
Scientific offers complete telescope kits that include all parts and detailed instructions for assembly. Their mirrors are made of Pyrex brand heat-
resistant glass. A mirror that has been ground at the
factory will have a higher price tag than a mirror
grinding kit offered by Edmund Scientific. A 6-inch

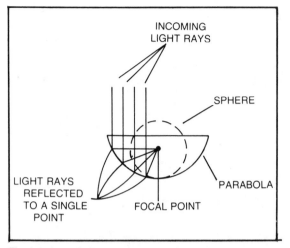

Fig. 9-1. A parabolic mirror produces the best possible image in a reflecting telescope.

INCOMING LIGHT RAYS

SPHERE

PARABOLA

LIGHT RAYS REFLECTED TO A SINGLE POINT

FOCAL POINT

primary mirror in Edmund's catalog will cost approximately $90, while a 6-inch primary mirror grinding kit costs about $40.

Edmund's kit includes everything necessary to perform the grinding and polishing process. It contains a mirror blank, plate glass tool for grinding, first surface rectangular mirror diagonal, sighting rings, magnifying lens for inspection, and eight abrasives including coarse to fine jeweler's rouge and tempered burgundy pitch. The only thing not included is the Foucault tester, which is offered separately in kit form. The Foucault tester kit can be bought for around $15.

WORK AREA AND MOUNTING PLATFORM

You need a working area that is as dust-free and clean as possible. The piece of glass to be ground is very sensitive to dust, which may cause scratches and otherwise damage the surface. The room temperature should be kept as constant as possible, because variations in temperature may cause difficulties. You need some type of mounting for the grinding tool, which is actually a second piece of glass of the same diameter as the piece to

be ground. The grinding tool is usually a little bit thinner than the piece of glass that will be the mirror.

The stand, table, or other platform should be at a convenient height that will make the grinding process as comfortable as possible. The mounting platform should provide access from all sides to enable you to change the direction of your strokes periodically to obtain an accurate curve.

ROUGH GRINDING

Examine the blank (the piece of glass to be ground) quite carefully to make sure it is not defective. The edges are probably a bit rough. These rough edges should be smoothed before rough grinding the blank. There is a greater chance of chipping, particularly in the finer stages of grinding when the curve in the glass is increased. To smooth the edges, use a medium-grade Carborundum stone. Both the stone and the blank should be dampened slightly before performing this procedure.

Place the blank on a flat surface and hold it firmly with one hand. Apply the Carborundum stone to an edge and begin smoothing it. Use a stroke in a direction away from the center of the blank. Apply a smooth, even stroke and slowly rotate the blank so that all edges receive the same amount of pressure. Do not apply too much pressure, as you may chip the mirror.

The tool is secured on the mounting platform after the edges have been smoothed (Fig. 9-2). Three cleats are secured in the mounting platform by means of screws 120 degrees apart. Make sure that the cleats and screws are below the surface of the tool, so the mirror will not come in contact with them when it is being ground. There should be a *small* amount of play so that it is convenient to remove the tool, but this should be a bit less than 1/32 inch.

Flannel or a few paper towels should be placed on the platform first. These will hold the tool firmly in place during the various grinding stages.

Abrasives are graded as to their coarseness. The coarsest ones are used in the rough grinding stages. The finest abrasives are used in the final grinding stages. Clean the mirror, tool, working area, and your hands after using each abrasive. The surface can be damaged or scratched quite easily if the wrong abrasive is used at the wrong time.

A number 80 Carborundum abrasive is usually used for rough grinding. The higher the number identifying the particular abrasive, the finer it is. Sprinkle a small amount of water on the tool. Apply a thin layer of the number 80 Carborundum.

Grinding will require a smooth and steady motion while rotating your position around the mounting platform. The mirror's center should travel over chords or lines on the tool in order to produce a hollowing action (Fig. 9-3). The center of the mirror is ground out by a motion that brings its center close to the edges of the tool.

Proceed along one chord in this manner for approximately 10 strokes. Take a new position in either direction while also turning the mirror slightly in the opposite direction to create a different line. These movements should be quite small and should not be done very rapidly. Some pressure

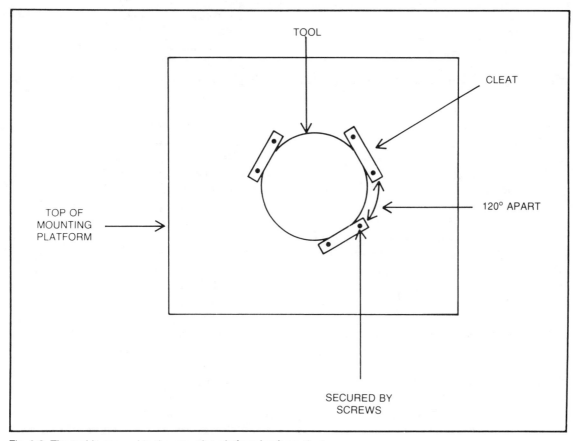

Fig. 9-2. The tool is secured to the mounting platform by three cleats.

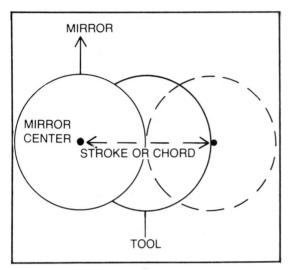

Fig. 9-3. The mirror's center should travel over chords.

will be required for the number 80 Carborundum abrasive to perform properly. A loud grinding noise will be heard if the right amount of pressure is used. If the Carborundum abrasive shoots out from the edges as the two surfaces are ground, there is too much water being used in relation to the amount of abrasive. You must continue adding abrasive during grinding. Remove all traces by rinsing with a sponge or cloth before applying an additional amount. As the abrasive's action is decreased , it will become a bit muddy. The grating sound will diminish considerably, indicating that it is time to add more abrasive. One period of time during which the abrasive is ground down to where its grinding action is limited is a *wet*.

After a while the tool will begin to have a hump, while the mirror will present a hollowed-out appearance. Exact and precise movements are not all that critical during the rough grinding process. The opposite is actually true. If there is a bit of variation from complete preciseness throughout, each variation will tend to compensate for possible errors and maintain a certain amount of uniformity in the curve.

The amount of time required for this stage of the rough grinding will vary considerably. Because of the amount of pressure required, you probably will not be able to complete the rough grinding in one session. Pick up where you left off and establish the same basic grinding pattern again. The time required for this stage may be anywhere from 3 to 6 hours.

After completing a full circle around the mounting platform, you can perform the first test on the curve. Most kits will include a template for this purpose and complete directions explaining how to perform the test. You will be checking the radius of curvature of the mirror, which will tell you if the curve produced during rough grinding is satisfactory regarding the intended focal length and the diameter of the mirror. Exact measurements at this point are not absolutely necessary, because further grinding is going to take place in any case. This test is simply to indicate whether the curve is taking form properly up to this point.

Use less pressure to complete the rough grinding. The strokes will be designed to compensate for the inaccuracies that have been determined with the template. In this stage the strokes should be a bit shorter and gentler, which will effectively correct those inaccuracies discovered during the test with the template.

FINE GRINDING

When you have to change abrasives, the work area and everything near it must be carefully cleaned to remove any particles of the last abrasive used. This cannot be overemphasized, because any larger particles remaining from a coarser abrasive can cause scratches to the mirrors' surface. The first step then must be repeated.

The purpose of the fine grinding stage is mainly to smooth out the mirror's surface. The stroke to be used is in the shape of a "W" with the motion centered over the tool (Fig. 9-4). Very little pressure is used at this stage. The abrasive used at this stage is the number 180 Carborundum, and it is applied in the same manner with a sprinkling of water. The mirror and the tool should change places at every other wet during fine grinding to make sure that the edges of the mirror are not ground too much.

You may spend approximately 1 hour with each type of abrasive. This is only an approximation. To determine when it is time to discontinue with one abrasive and begin with a finer one, observe the pocks or pits on the mirrors' surface left by the previous abrasive. When they appear to be disappearing uniformly, switch to a finer abrasive. This inspection will be easily accomplished by holding the mirror up to an artificial light.

The chances of scratches occurring are now better. One way to prevent damage is to change the abrasive mixture more frequently. Don't allow it to dissipate too much during a single wet. This is also the point where it is dangerous for any of the edges to chip or crack. To avoid any scratches, the tool, mirror, and work area should be cleaned thoroughly after each wet, even if the same abrasive is being used.

POLISHING THE MIRROR

Apply a layer of pitch to the surface of the tool. Because the polishing process will depend on air circulation, the use of water, and polishing materials, the pitch lap must be constructed properly. It is not within the scope of this book to explain how a pitch lap is made. Any kit will contain complete instructions. Figure 9-5 shows a completed pitch lap as it should appear on the surface of the tool.

The mirror is placed over the pitch lap and firmly pressed down so that the pitch lap takes the exact form of the mirror. The whole process will

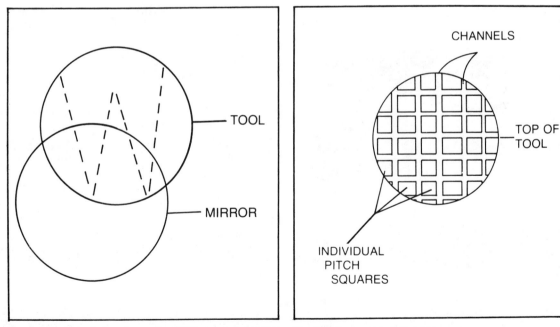

Fig. 9-4. A stroke in the shape of a "W" is used during fine grinding.

Fig. 9-5. A completed pitch lap.

probably take a full day's work, including the time that must be allowed for the pitch to be heated, cooled, cut, and installed on the tool. The actual pressing of the mirror down on top of the tool should be done for about 1 hour before polishing is begun.

The strokes must be quite smooth and consistent during the polishing of the mirror. It will make little difference whether the mirror is on top or vice versa. The rouge supplied with the kit is mixed with an amount of water specified. The rouge is usually applied with a small brush, and the wets should be of short duration. Polish the mirror for a long time, as this will produce the best results.

Because the normal polishing time for a 6-inch mirror will probably be around 6 to 8 hours (actual polishing time), the lap will probably need to be maintained at intervals so that it keeps the proper shape. This is done periodically with a measured stick inserted in between the squares. A chisel is then inserted to chip off the uneven parts of the

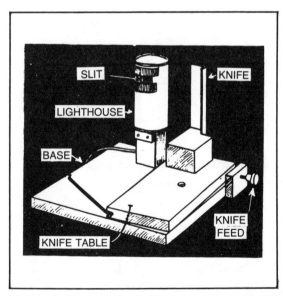

Fig. 9-6. The Edmund Scientific Foucault tester is designed to accurately test the curve of the mirror (courtesy Edmund Scientific Company).

edges of each square. Remove any chipped pieces from the channels.

THE FOUCAULT TESTER

The Edmund Scientific Foucault tester costs less than $15 and will accurately test your mirror's curve (Fig. 9-6). The tester is used to detect any bumps or inaccuracies in the mirror's surface. A beam of light from a source at the radius of curvature, or double the focal point, is directed at the mirror's surface. If the surface is perfectly spherical, the light beams will be radiated back to the exact point from which they came. If a slice is made in the beam at a point in the line of the image, it will strike a shadow. If you are behind the beam, you will be able to see any irregularities in the mirror's surface. Refer to the instructions contained in the tester's kit for more information on its use and specific procedures for any inaccuracies pointed out by the test. These inaccuracies can usually be corrected by several polishing variations and adjustments to the pitch lap.

PARABOLIZING THE MIRROR

The grinding process has produced a sphere. This shape must now be a parabola. The *classical method* is the simplest way to parabolize a mirror. The mirror is placed on top of the pitch lap. Very long strokes are used for short periods—approximately 10 minutes at a time. There should be intervals separating the periods to give the mirror and pitch lap time to stabilize.

COATING THE MIRROR

When your piece of glass has been polished, parabolized, and tested for its accuracy, it is now time to transform it into a mirror. Coat the glass with a thin layer of a light-reflecting metal. Earlier mirrors and optical devices were coated with silver. Aluminum is the most popular choice today. It is relatively inexpensive and produces aluminum oxide, which protects the surface from deterioration for many years. Because the process involved in aluminizing a mirror is quite complicated and requires specialized equipment, it should be done by a professional. There are alternatives to aluminizing, but these are time-consuming and require you to mix chemicals.

Chapter 10

The Home Observatory

T HE IDEAL HOME OBSERVATORY IS AN ENCLOSURE that can be periodically opened to view the heavens while the astronomer remains in relative comfort—free from the perils of the environment. It must be large enough to house the telescope, accessories, and the astronomer.

The classic domed observatory revolves by motor-controlled devices with the earth's rotation. Companies specialize in building these tiny observatories. Astronomy magazines have advertisements for complete observatories or specialized domes that can be fitted to a small cinder block structure. These domes are not usually motor-controlled. They must be rotated by hand to accurately track a long-time exposure photography attempt.

The astronomical dome is probably the best bet when it comes to comfortable, quasi-indoor viewing. Most people cannot afford them, though.

One of the best low-budget observatories is a tent. Many tents are large enough to house a fairly large telescope. They certainly are enclosures. Electric power to drive the clock mechanism for automatic tracking can be supplied to a backyard installation by an extension cord run from the house supply.

The tent is not open to the sky and can't be rotated, but alterations can be made. A tent is inexpensive. When the alterations are made, you can pack your personal observatory in your car's trunk, along with your telescope and accessories, and search for the ideal viewing location. A power inverter can be bought that will allow you to run small appliances off your car's battery from a separate battery just outside the tent.

TENT ALTERATIONS

You must choose an appropriate tent for your

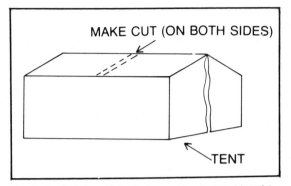

Fig. 10-1. A hole is cut laterally across the sloping sides of the tent ceiling.

telescope. Reflector telescopes must be mounted near the ground, as you must peer through the eyepiece at the far end. If you use a reflector telescope, choose a tent whose topmost section is relatively low so that the end of the telescope may be inserted. If your telescope is a refractor, a high ceiling is best. It will be necessary for you to peer through the eyepiece at the bottom end of the instrument. Refractor telescopes are best for a tent observatory.

After you have chosen the appropriate tent, the modifications are relatively simple. You need to cut a hole in the roof. Figure 10-1 shows how a lateral cut is made across the sloping sides of the tent ceiling. Use good cutting shears. Be careful to make a straight cut. Mark off a cutting line before starting. Two waterproof zippers are installed after the cut has been made. These are available from many camping supply outlets. Two zippers are used instead of one, so it will not be necessary to open both sides of the tent at the same time. These sides will be 180 degrees opposed, so this will not necessitate exposing the west side when the telescope is aimed toward the east.

The zipper pull tabs should ideally be located inside the tent, zipping from the bottommost side to the apex of the ceiling. Make certain that all stitching is waterproofed.

String up the tent loosely so there is considerable play in the ceiling. When the end of the telescope tube is inserted, the play of the fabric should not interfere with the telescope's tracking. The material will give as the scope tube rotates.

If constructed as shown, you will have to open the zipper completely for viewing overhead. If you install the zipper with the pull tabs at the ceiling when in the closed position, you only need to partially open the slit for overhead viewing. This can often lead to problems in waterproofing. For the colder climates you may want to install several zippers along this lateral cut, opening only the one that will give you the proper exposure. Another method involves installing a two-way zipper containing two pull tabs. A single zipper can be opened from either end using this arrangement.

Move the telescope inside and open the tent to expose the portion of the sky you want to observe. The arrangement of the opening will allow for great flexibility when viewing in any direction. If your tent is erected with sides pointing east and west, the telescope can be mounted in the exact center of the enclosed area for east-west viewing. When aiming the telescope north or south, its tripod must be moved around to allow the scope to be inserted through the slitted opening at a 90-degree angle. Figure 10-2 shows a standard setup for east-west viewing, while Fig. 10-3 shows the arrangement for north-south viewing using the same tent.

STORAGE BUILDINGS

Aluminum and steel storage buildings are available through most mail-order catalogs and from a hardware store. These buildings are designed for backyard installations and are meant to house things like power mowers and other lawn care implements. They can also be used for permanent home observatories. Modification will be required to make the roof movable.

Many structures are built around tubular steel

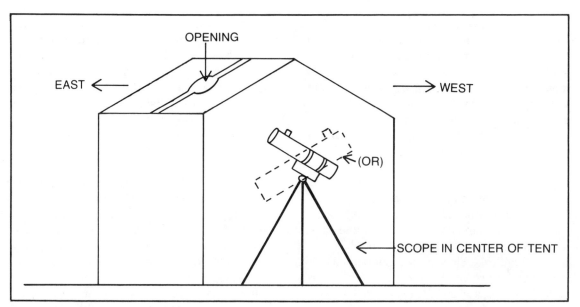

Fig. 10-2. A setup for east-west viewing.

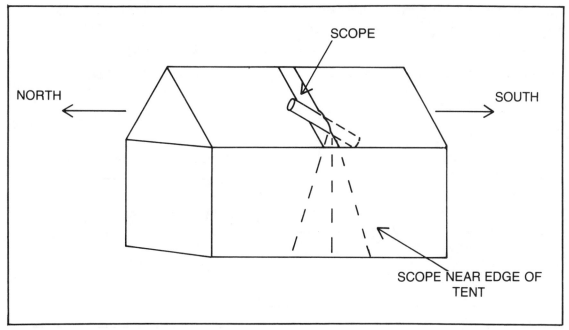

Fig. 10-3. A north-south viewing arrangement.

frames. The frame is first erected. The roof and siding panels are then attached. Modification involves cutting away panel sections of the roof, installing them on hinges, and then applying rubber molding to the edges to make these incisions waterproof. The hinges should be installed inside so that the panels can be pulled downward.

Reflector telescopes require small buildings with low roofs. Refractor telescopes should be used with slightly larger buildings that have more ceiling-to-floor height. Figure 10-4 shows how a typical arrangement might be handled.

The main problem with this arrangement is that the metal roof is not flexible like the canvas material in the tent observatory. If your building is installed with the sloping roof pointing east and west, most viewing will be limited to these east-west directions, as well as to overhead. If you want to look toward the northern horizon or to the south, cut other panels in the north and south walls. Because the building is constructed around a tubular frame, this cutting should not affect the overall

structural stability of the building. As more panels are cut, waterproofing the structure becomes a major problem.

Some buildings have a framework for the walls only, while the roof is a separate structure that is bolted on after the walls have been raised. These buildings can make very adequate observatories with some major modifications.

The walls are constructed in the normal manner, but the roof is not permanently bolted to them. It is simply laid in place so it will slide back and forth as desired. This requires additional support at the longitudinal ends of the building (Fig. 10-5). Choose small peaked-roof buildings. The roofs should be very lightweight. Make sure the roof edges have a few inches of overhang to allow rain to drain away from the wall edges.

It is very difficult to make this type of arrangement completely waterproof, although rubber molding can be installed along the longitudinal lines where the roof edges meet the walls. When rain is falling straight down, this arrangement is usually

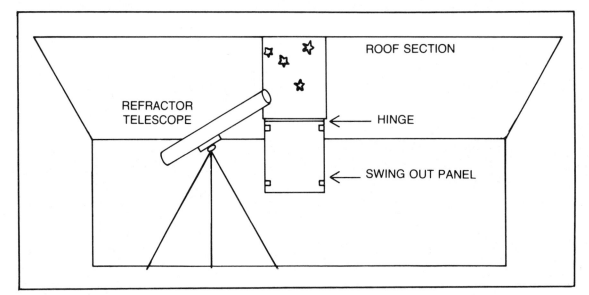

Fig. 10-4. A typical arrangement using a storage building.

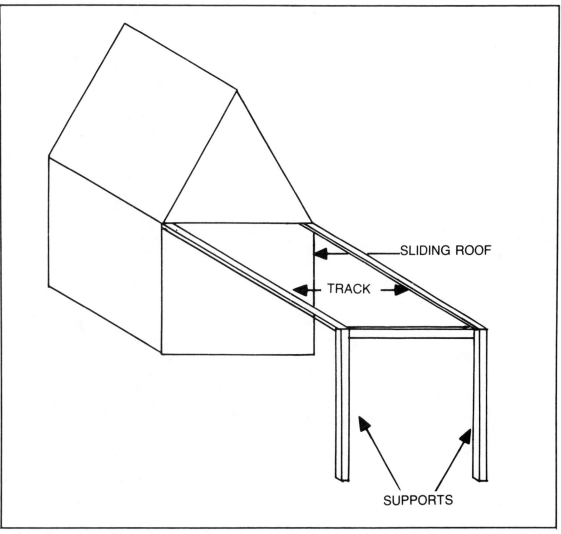

Fig. 10-5. Additional support may be necessary if the roof is not permanently attached to the sides.

high and dry. When this moisture is blown laterally by high winds, however, some seepage can occur without proper waterproofing.

Some buildings will require very little modification for the roof to slide back and forth. Others may need a track and roller system. This can be done inexpensively, but it takes lots of time. Figure 10-6 shows how casters are installed at four or five points along the roof edge and fitted into a steel track that can be purchased from your building supply outlet. The track is extended past the building on either side and supported by wooden posts or steel tubing.

To use this structure as a home observatory,

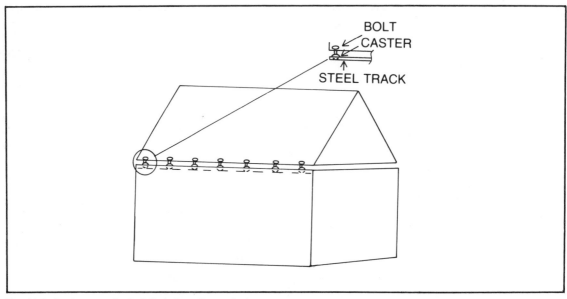

Fig. 10-6. Casters may be installed along the roof edge.

the telescope is mounted on the inside. The roof is moved back and forth. This allows 360-degree sky coverage regardless of the direction in which the building is pointing.

One major disadvantage of this structure is most of the building's interior must be exposed when viewing objects directly overhead. The roof must be pushed back to uncover a little more than half of the interior, assuming that your telescope is installed in the building's exact center. The scope may be moved to one end or the other, and a small portion of the roof can be pushed back.

East-west viewing for this type of building with its ends pointing north and south is an even greater problem. This will depend on the height of the building's walls. Movement of the telescope within the building can sometimes overcome these problems. Alternately, you may want to combine the last two observatory ideas and cut pull-away panels in the sliding roof. This should allow permanent mounting of the telescope in the building's center. You will have reasonable access to all parts

of the sky without exposing yourself and your telescope to the elements.

You may want to construct one of the many wooden buildings featured in various how-to-build-it magazines. Money can be saved. The design can be treated as an observatory from the start, with the plans being modified to suit individual preferences.

If you want to use a garage or other building already on your property, this is fine, too. You will probably have to build a platform inside any of these structures to have a means of getting the scope nearer to the roof area. This can be done quite easily and for a fraction of the cost of even the smallest storage building. Slide-out panels can be cut in the roof surface and insulated as previously described.

A skylight may be installed. Some skylights are domed structures, while others consist of flat plates of glass. Skylights can often increase the value of a home, but they are quite expensive to install. Most of the expense involves the work required to make the installation absolutely waterproof. The skylight must be made removable or slidable for astronomy uses. You cannot get good viewing results when peering through a plate glass window.

One amateur astronomer installed a do-it-yourself skylight (of sorts) in the side of his peaked roof. His observatory was his attic. The skylight was really a set of sliding glass doors installed on a diagonal plane instead of vertically. When the installation was complete, he merely opened one of the doors to a point where a portion of the sky to be viewed was exposed. The main drawback was that only 45 percent of the sky was exposed by this method. When the object being viewed passed through a point directly overhead, it was lost until the following night.

A simpler installation might involve putting in a large window that can be raised and lowered as desired. If you go this route, check with the local building contractor to see how this can best be accomplished. Ask him to pay special attention to the waterproofing problems that will certainly be encountered.

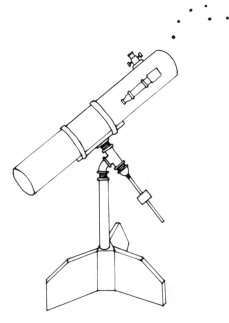

Chapter 11

Telescope Care, Maintenance, and Adjustment

Y OU MUST LEARN HOW TO CARE FOR AND MAIN- tain your telescope. A malfunctioning tele- scope will produce inaccuracies during observa- tions that may result in incorrect conclusions. Some distortions caused by faulty alignment and other malfunctions can be readily apparent to an experi- enced user of telescopes, but the amateur may be confused by these problems. Most of the proce- dures discussed in this chapter are very simple to perform and do not take a very long time.

Although a telescope is a fairly durable instru- ment, it cannot withstand being mishandled, dropped or left exposed to the elements. A tele- scope that has been dropped may need to be realigned, because the optical system will be jarred. Some manufacturers provide for factory repairs in- volving alignment of their telescopes, but you may perform this procedure if the instructions provided in this chapter are followed carefully. The proce-

dure is collimation. Most manufacturers align their telescopes at their factories prior to shipment. The only time that a telescope will need to be recolli- mated is if it is dropped, jarred severely, or under- goes sustained jostling. Before examining collima- tion, though, we will discuss general telescope cleaning and care. Information from Cross Optics pertaining to their Newtonian reflector, the Nustar 14, has been used.

CLEANING MIRRORS AND LENSES

A thick coating of dust on one or more optical surfaces in a telescope can reduce image contrast by more than 25 percent. Many products available from camera stores will effectively remove dust from a lens. A 1-inch brush designed for use with cameras and telescopes will eliminate the static charge that can build up on glass. A static charge will draw dust particles of opposite electrical charge as well, if not better, than a magnet will draw steel nails.

The cleaning method described here is simple, effective, and safe for the optics of your telescope if done carefully. You can use this procedure for cleaning eyepieces or other lenses. Purchase a box of sterile cotton that comes in rolls—not balls. Wash your hands thoroughly to remove most of the natural oils that have accumulated. Tear off a piece of cotton, remove any abrasive particles, and exhale on the lens surface you wish to clean. Gently swirl the cotton around on the lens surface to wipe off the moisture that condensed from your breath. Make sure you produce a good "fog" of moisture before expecting the oil film or whatever you are cleaning off to budge. Do not let any of the moisture get between the lens and barrel, as this may eventually cause the cement on the elements to separate. This procedure may be repeated as necessary to adequately remove an oil film (such as from your eyelashes hitting the eye lens).

For really stubborn smudges or oil films, acetone or isopropyl alcohol applied to a piece of cotton can be used with satisfactory results on coated lenses and mirrors. The chances of ever having to take individual lenses apart to clean them are very slim. Only the exposed lenses are likely to ever need cleaning. You can use this procedure when cleaning the correcting lens of a catadioptric telescope.

Cleaning mirrors is very much like cleaning lenses, except it is first necessary to remove the mirror from its cell and submerge it in a diluted solution of water and detergent before proceeding to wipe the surface. The container for the water and detergent solution must be free from any other chemicals or debris, as this may permanently damage the mirrors. Wipe mirrors carefully. Be very gentle to avoid scratching or sleeking the metal coating on the glass. As with lenses, any wiping should be done with a small piece of sterile cotton. Small bits of foreign matter stuck to the primary mirror's surface will actually have little effect on the performance, whereas overzealous rubbing or scrubbing may wear the coating thin.

The cleaning solution should consist of a non-film producing gentle detergent at dilutions of 200:1 to 300:1. Any gentle liquid will work; milder ones produce the best results. If the foreign matter will not come off with this solution, try using the same solution in a less diluted form. Rinse with running tap water, then rinse again with a small amount of distilled water. To remove the last few water spots, you may spray the mirror with ether. This last step is not absolutely necessary. Stand the mirror on its edge and spray the surface until all the water droplets are removed. Use caution when using this flammable substance. Avoid smoking, any open flames, or any other heat source during this procedure. Provide sufficient ventilation.

The secondary mirror's condition is more critical per unit area than that of the primary mirror. A dirty surface or thin coating of dust reduces the effective light-gathering power and contrast on a degraded surface. You might have the secondary mirror coated with enhanced silver coatings that reflect 97 to 99 percent of the incident light. Keep special dustproof cover over the secondary mirror when the telescope is not in use. This dust cover must not touch the surface of the coating.

Perhaps the most complete cleaning of an optical surface short of doing an ion glow discharge cleaning in a vacuum bell jar is achieved by using the collodion technique of mirror cleaning. Surfaces cleaned by this method have perhaps 10 times less residual contaminant and particulates as compared to a methanol-distilled water-cleaned surface. The technique was first reported by James B. McDaniel in *Applied Optics*, Volume 3, No. 1, January 1964, on pages 152 and 153. John B. Tyndall produced a concise report (and an improvement) on the technique that appeared as National Aeronautics and Space Administration (NASA) Tech Brief 70-

10463, August 1970, from which the following quotation has been extracted:

"A modified cleaning method of a collodion cleaning technique is being utilized whereby the mirror can be cleaned in its holder, saving manpower, system down-time, and minimizing mirror damage. The modification consists of the addition of a layer of cheesecloth embedded in the collodion, which aids in peeling the dry collodion coating from the mirror. The only equipment necessary is collodion, cheesecloth, and a soft camel's hair brush. Only U.S.P. collodion (cellulose nitrate in ether-methanol solution) should be used. 'Flexible' collodion with a camphor plasticizer cannot be substituted.

"A thin even film of collodion is painted over the entire surface of the mirror using the camel's hair brush. It may be desirable to scrub the mirror very lightly with collodion using the brush. Application of five or six thin coats should be sufficient. A few minutes should be allowed for each coat to dry, lending support to the following coat without running. Dryness can be determined by pressing the finger gently against the mirror surface. The entire mirror surface should be inspected to insure that all areas are gelled and dry. After the first three or four coats have completely dried, two layers of cheesecloth should be placed over the mirror with a 2 or 3-inch overlap beyond the mirror edges. One or two additional coats of collodion should be applied over the cheesecloth to allow seepage through to the original coats, making a solid seal of cheesecloth and collodion over the entire mirror surface. After sufficient time to dry, the overlap should be peeled off the mirror by a gentle and even pull on the cheesecloth. The mirror should be clean and ready for use. If not, the process should be repeated. Mirrors having severe damage, however, cannot be effectively cleaned by this process and must be sent to the factory for recoating.

Image turbulence is much reduced or near zero on evenings of heavy dew fall or frost. In order to observe under these conditions, the use of either a dew cap (short cardboard or plastic tube whose length is about 1-1½ times the clear aperture stuck on the skyward end of the telescope tube) or a hair dryer, with which you can evaporate the dew or frost, is a necessity. Note that the hair dryer cannot be powered through the ac/dc converter of a telescope's frequency generator-slow motion. The hair dryer draws too many amps. Note that only a brief blast is necessary. Too much hot air will cause undesirable convection currents. Dew forms less readily on clean optics compared to dirty optics, because particulates that act as condensation nuclei are not present on clean optics.

A good aluminum coating will be quite serviceably reflective for about five years in an open-tubed telescope. A thin film of dust usually looks more harmful (when examined with a flashlight from the front of the telescope's tube) than it really is. Overly zealous or too-frequent washings (more often than every three or four months) may require you to have your mirror aluminized every one to two years. Keep your telescope's optics capped with dust covers when not in use.

COLLIMATION OF THE CELESTRON 14

Collimation is the technique of aligning the optics of a telescope. This is a relatively simple procedure. For our discussion on collimation, the Celestron 14 telescope is used as an example.

Collimation simply means that the optical centers of the optical elements are square with each other or perpendicular to the optical axis. The only collimation adjustment necessary or possible with the Celestron 14 telescope is the tilt adjustment of the secondary mirror.

You will need a proper light source to check collimation. A bright star near the zenith is best (to minimize atmospheric scintillation), but Polaris will do. Your telescope should be in thermal equilibrium with its surroundings during collimation. If the instrument is transported between very great temperature extremes, allow about 45 minutes for it to reach equilibrium.

Using a 25-mm eyepiece, defocus the telescope so that the out of focus blur circle of your light source occupies about a third of the field of view. If the shadow of the central obstruction (secondary housing) is not perfectly centered inside the blur circle, the telescope is out of collimation. Even if the shadow appears centered at this point, continue to read the techniques used for collimation.

To adjust the collimation, use the slow-motion controls to repoint the telescope so that the blur circle is moved to the edge of the field in the direction that the shadow is off-center. Bring the blur circle back to the center of the field using the three Allen screws at the edge of the secondary housing (Fig. 11-1).

Tighten the screw(s) in the direction that the shadow is off-center and loosen the other screw(s). Tighten the screw(s) so they are only finger-tight.

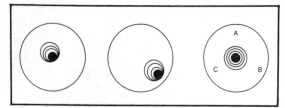

Fig. 11-1. The step-by-step method of collimation (courtesy Celestron International, Torrance, California, U.S.A.).

Repeat this process until the blur circle is again at the center of the field. With the blur circle again centered in the field, the shadow of the central obstruction may still be a bit off-center. Repeat the collimation process until the shadow is perfectly centered within the circle.

Repeat the collimation process as necessary using successively higher-powered oculars until you reach the highest powered ocular that you will be using. Collimation at the higher powers (6 mm and up) is best accomplished with the telescope in focus. When collimating in focus, you will be observing the Airy disk instead of the shadow of the central obstruction. This will appear as a bright ball with a single diffraction ring around it. When the ball is exactly centered inside the ring, your telescope is collimated. These basic collimation adjustments will apply to most Schmidt-Cassegrain telescopes.

COLLIMATION OF A REFLECTOR TELESCOPE

Most amateur astronomers elect to build reflecting telescopes. This is a relatively easy task when proper adjustment points are provided at the primary and secondary mirrors.

Begin by aiming the opening of the telescope toward a source of fairly bright light. Collimation will best take place in the daytime when a bright sky can be used as this source. Remove the lens from its holder and stare down the opening at the secondary or diagonal mirror. Your eye should be clearly seen centered in the diagonal. You should also see something similar to Fig. 11-2 if both mirrors are in proper alignment. Some inexpensive telescopes come with the diagonal already adjusted, and no means has been made for further movement. More expensive instruments will also require alignment of the diagonal.

Start with the primary mirror alignment. If this is incorrectly adjusted, you will see something similar to Fig. 11-3. This shows the tube opening and a portion of the inside of the tube assembly. When the primary mirror is aligned properly, the bright light at the opening will appear to be centered as shown earlier. Adjust the correcting screws on the primary mirror while noting which direction the aperture seems to take. You will bring the opening into the direct center of the aperture you are peering through by careful alignment.

If the diagonal mirror is out of alignment when the primary mirror is properly adjusted, you may see something similar to Fig. 11-4. Make adjust-

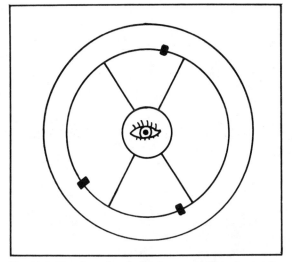

Fig. 11-2. The view when both mirrors are aligned properly.

cated here is simple and will give excellent results.

After a preliminary adjustment has been made, you may want to complete a precision alignment. This is done at night by sighting the telescope on a bright star lying directly overhead. This is a trial-and-error method and may require an assistant to make adjustments to the primary mirror. Use a very low-powered lens and defocus it until you begin to get a ring pattern (Fig. 11-5). This indicates perfect alignment with each inner circle falling at the exact center of the outer one. If there is a slight centering error, the image may appear more similar to the one in Fig. 11-6. When this occurs, make some slight adjustments to the primary mirror until the correct image is received. For finer adjustments, it will be necessary to use a medium to high-powered eye-piece.

Point the telescope to the moon or a planet such as Saturn or Jupiter. Look for any unusual appearances in the image received at the eyepiece.

Precise alignment requires exceptionally good astronomical viewing conditions. Don't attempt it on a night when there is considerable atmospheric turbulence or any thin cloud cover. These adjustments are best begun in rural areas whose skies are relatively free of atmospheric contaminants.

ments to the primary mirror mounting until it lies within the exact center of the aperture.

This type of alignment is known as coarse or preliminary adjustment. The entire procedure can usually be accomplished in a few minutes and aligns the two mirrors fairly accurately. Many types of alignment instruments are advertised, but most are unnecessary expenditures. The procedure indi-

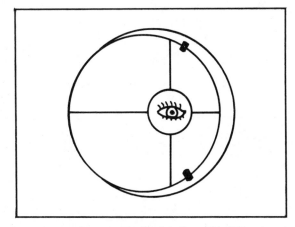

Fig. 11-3. The image will appear similar to this if the primary mirror is not properly aligned.

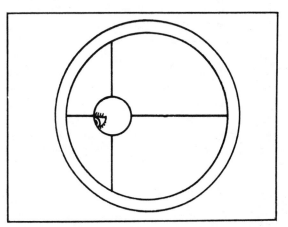

Fig. 11-4. The diagonal mirror is not aligned properly.

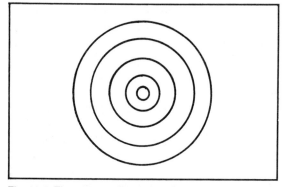

Fig. 11-5. The pattern will look like this in perfect alignment.

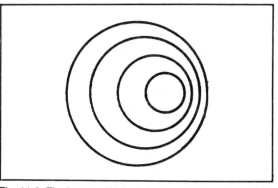

Fig. 11-6. The image will be similar to this if there is a slight centering error.

The precise alignment procedure is not often required for amateur astronomy work. The coarse adjustment will usually suffice. If you build a telescope, you will want to go through a complete coarse and precision alignment before first use. The instrument should hold this alignment for some time if adequate care is taken to avoid high impact shocks to the tube assembly.

OPTICAL CARE OF A DRIVE CORRECTOR

A drive corrector is an attachment to a telescope that makes it possible for deep sky photographers to correct for image drift in right ascension by speeding or slowing the telescope's motor. If the inside of one of these units needs cleaning, be careful when opening the corrector. The corrector plate is quite thin and may be damaged. This plate is very thin and must be replaced in exactly the same orientation as it was prior to removal. This is for collimation reasons and also because both corrector and secondary mirror are position-matched for optical performance with respect to the primary mirror.

The information provided by Celestron regarding their Celestron 14 telescope is used here. To remove the corrector lens, unscrew the eight Allen-head setscrews and remove the corrector re-

taining ring. After removing the corrector retainer, a code number etched onto the corrector's edge and some cork shims between the edge of the corrector and the front-cell ledge will be visible. These shims protect the corrector from shock and hold its optical center over the optical axis of the scope.

Before removing the corrector, pencil index marks on the inside of the front cell that indicate the precise positions of the corrector code number and each shim. Remove the number-code the shims so they may be returned to the same exact positions when the procedure is completed.

Grasp the secondary mirror cell and lift the corrector out of the tube for cleaning. Because the secondary mirror is mounted in the aluminum cell in the corrector's center, it will also be removed by this same procedure. If necessary, the secondary mirror may be removed by unscrewing the center screw on the front of the secondary mirror cell. When the secondary mirror is remounted, the index line on the mirror's back must be pointing to the center of the code number etched on the corrector lens.

To clean the corrector lens, use a can of pressurized air or a camel's hair brush. A photographic lens cleaner may be used with white Kleenex tissue or a nonsilicone photographic lens tissue to clean the corrector. Do not clean the corrector with vigorous circular motions. Use several tissues. Take a single gentle wipe from the center out with each tissue.

When replacing the corrector, align its code number with the index mark made on the tube and return each shim to its proper position. When replacing the corrector retainer, tighten the screws down gradually, in round-robin fashion, to finger-tight only. This should be just barely tight enough to keep the corrector from moving whenever the telescope is repositioned. Also, too much tightening may cause the corrector lens to crack.

Most manufacturers will provide complete instructions on handling, care, and general maintenance of their telescopes and optional equipment. If there is any variation between the information provided here and that of a given manufacturer, adhere to their instructions. The procedure itself is not difficult. Take care that the components are reinstalled exactly as they were before being removed. These parts are somewhat fragile and may be damaged if handled in a rough manner.

GENERAL CARE OF YOUR TELESCOPE AND ACCESSORIES

The body of the telescope should receive a periodic cleaning. Dust and dirt buildup on your telescope's body may eventually blur viewing or cause permanent damage to the exterior finish. Neglecting your telescope's appearance will reduce its resale value.

Be careful when transporting the telescope to an observing site away from your permanent mounting. Avoid jarring or dropping the instrument, as this may distort the optical alignment and cause unsightly scratches on the body. If transporting the telescope in a vehicle, provide soft cushioning for the instrument and a means to brace it.

Don't put any telescope parts to be assembled on the ground. If the area is damp or the ground is somewhat muddy or even gravelly, damage will most certainly result. Most telescopes are easy to assemble at remote locations and can be made operational quickly. Larger telescopes consist of three or four large components that can be transported and assembled easily. Strict adherence to each manufacturer's instructions should be followed.

If the body of your telescope becomes dirty or appears to be losing its original luster, it should be wiped with a clean damp cloth. Never use any type of cleanser or abrasive product. The cloth to be used should be kept separate from other cleaning cloths.

When your telescope is not in use, it should be stored in an area where it will not be exposed to the elements. It is not designed to withstand severe temperature changes, weather variations, etc. Any type of condensation may cause internal damage. The telescope and its accessories should be stored in an area of medium to low humidity to avoid condensation. Make sure the appropriate protective covering is provided for any lenses.

Chapter 12

Astrophotography

A STROPHOTOGRAPHY, A PROCESS OF PHOTO-graphing what is seen through the telescope, is becoming extremely popular. Relatively inexpensive and yet sophisticated astrophotography equipment can be attached to most standard telescopes.

Astrophotography allows you to photograph what is seen through the eyepiece of the telescope and to display these scenes, when development has been completed, for all to see. A photographic image can be enlarged many times and often will reveal details that are not visible during the initial viewing process. Direct viewing of the skies often means limited observation time due to atmospheric conditions, ambient light, and approaching daylight. Complex measurements of celestial bodies may be taken by examining a series of photographs from film that has been exposed at specific time intervals.

Another advantage of photographic images of the sky is that longer exposures can be utilized. Objects that may appear as mere faint images through the telescope will appear in greater clarity. It is possible to accurately identify a celestial body as to its classification and orbital path. This is best done with a series of 10 photographs of the same area of the sky at regularly spaced intervals. This method has been used in the past to study our sister planets and their satellites.

EQUIPMENT

Good astrophotography takes time to learn. It should be undertaken only after you have mastered the use of your telescope.

Figure 12-1 shows a Celestron C-90 telescope equipped with an adapter for astrophotography. This effectively turns the instrument into a large telephoto lens that is coupled directly to the cam-

Fig. 12-1. The Celestron 90 telescope equipped with an adapter for astrophotography.

Fig. 12-2. The Celestron 90 equipped with a T-adapter attached to the eyepiece mount.

era. Two devices are used with the Celestron telescope for astrophotography work. Figure 12-2 shows a T-adapter that attaches directly to the telescope eyepiece mount. In addition to its coupling effects, the T-adapter separates the camera from the scope by the proper distance to achieve full frame image and correct focusing. Figure 12-3 shows the T-ring that attaches to the end of the T-adapter and to the standard camera lens mount. Because different 35-mm cameras require different lens base mounts, T-rings are made specifically for various cameras. A T-ring designed for attachment to a Pentax camera would not fit one made by Canon. You must order a T-ring built especially for the camera to be used.

For still photography with your 35-mm single lens reflex camera body using the Celestron telescope as an example, all you need are Celestron's T-mount camera adapter and a T-ring for your specific camera. The T-adapter couples the camera body to the rear cell of the telescope by means of the T-ring and permits the camera body to be oriented in either a vertical or horizontal format.

For greater mobility during photographic sessions or for motion picture photography, the tube assembly of the Celestron 5 or Celestron 8 telescope demounts from its fork for use on a photographic tripod. The tube is mounted on the tripod using one of Celestron's photo tripod adapters. These adapters will fit the standard ¼-20 tripod head, and the tube assemblies are available as separate items with the Celestron telescopes. The Celestron 5 and Celestron 8 tube assemblies include finderscope, tripod adapter, and the T-

Fig. 12-3. The T-ring is designed for attachment to the T-adapter and the camera lens mount.

adapter. The Celestron 14 tube assembly includes only the finderscope and T-adapter.

You may want increased subject contrast, color temperature conversion, or special filtration effects. Celestron offers the Series VI drop-in filter set shown in Fig. 12-4. The set consists of six ring-mounted, optical glass filters that fit into the rear-cell recess of the telescope ahead of the T-adapter.

When the element of motion is added to close-up photography, some very nice visual effects can be obtained, particularly in the realm of long-distance macrophotography. For motion picture photography with the Celestron telescopes, the company offers a special adapter that couples C-mount reflex cameras to the tube assembly by a T-mount camera adapter.

Celestron states that motion picture photography with their instruments requires a custom-fabricated mounting bar to couple the tube assembly rigidly to the camera body. The Beaulieu 4008 ZM4 (super 8-mm) and the Bolex EBM H16 (16-mm) are motion picture cameras used suc-

Fig. 12-5. The Celestron equatorial wedge assembly (courtesy Celestron International, Torrance, California, U.S.A.).

cessfully with Celestron telescopes. The drop-in filters for still photography will provide additional flexibility in making exposures during motion picture photography.

The simplest and probably least expensive form of astrophotography is lunar and planetary photography. It is an ideal way to learn the first steps of photographing the heavens. You can capture the belts of Jupiter or even the crater Copernicus in a photograph.

The simplest form of lunar or solar photography—full-disk or wide-field—employs the same camera coupling used for still photography. Simply mount the complete telescope on Celestron's equatorial wedge assembly and locked-triangle tripod (Figs. 12-5 and 12-6). The electric drive is turned on, and exposures of .006 second or so are all that is necessary.

For close-up lunar or solar photography or for planetary photography, the T-mount camera adapter is replaced with Celestron's *Tele-Extender* and visual back with an ocular used to project a magnified image to the film plane (Fig. 12-7). The

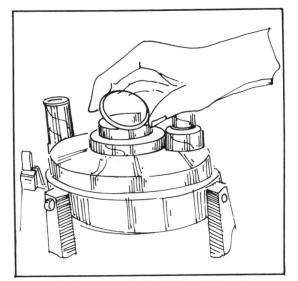

Fig. 12-4. The Celestron Series VI drop-in filter set.

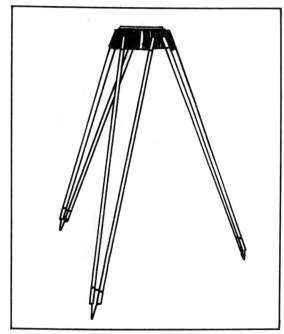

Fig. 12-6. The Celestron locked-triangle tripod.

Tele-Extender is shown mounted in Fig. 12-8. Select the amount of magnification that you desire by the focal length of the ocular that you use. The 40-mm ocular increases the image size by 3.4

Fig. 12-7. The Celestron Tele-Extender is for close-up lunar or solar photography (courtesy Celestron International, Torrance, California, U.S.A.).

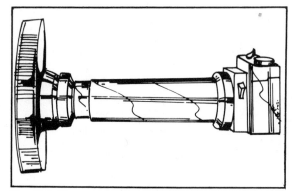

Fig. 12-8. The mounted Tele-Extender.

times. The 25-mm, 18-mm and 12-mm oculars increase the image size by 6, 8.7, and 13.6 times respectively. There is nothing gained by using shorter focal length oculars in the Tele-Extender, as the diffraction limit of the telescope would be exceeded.

Using this Tele-Extender with the Celestron 5 or 8 telescope places the camera back far enough that the out-of-balance condition could interfere with proper operation of the clock drive. A set of counterweights for each of these two instruments is offered by Celestron as an optional accessory to achieve proper balance. A Celestron telescope with the counterweights in place is shown in Fig. 12-9. Counterweights for the Celestron 14 are included with the instrument.

DEEP SKY PHOTOGRAPHY

Many time exposures of deep sky objects require *guiding*, a process whereby you make fine adjustments in telescope point during the course of the exposure. Guiding is necessary because even with an electric drive, there is a very slight movement of the image in the field of view. This movement cannot be tolerated on film. Celestron offers an optional ac-dc drive corrector for their telescopes and an illuminated-reticle ocular assembly (Fig. 12-10).

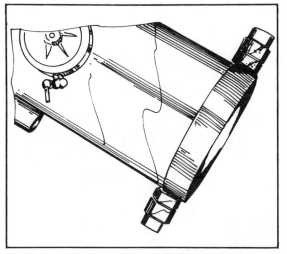

Fig. 12-9. A Celestron telescope with the counterweights in place.

The drive corrector speeds up or slows down the electric drive of the Celestron telescope for adjustments in celestial longitude. (Adjustments in latitude may be made with the declination slow-motion control.) Instant-response "fast" and "slow" buttons on a remote-control box provide for a 50 percent increase or decrease in drive speed. The corrector, which can either operate on 110-volt, 60-hertz household current or act as a dc inverter, also has an output jack for the illuminated-reticle ocular assembly. Adapter cables for battery and automobile cigarette lighter are also included. The drive corrector also has a calibration control that permits sustained vibrations in drive speed of up to plus or minus 10 percent—a useful feature when critical resolution is desired in lunar, solar, or planetary photography.

To determine just how much correction is needed during the course of an exposure, Celestron offers the illuminated-reticle ocular assembly. This is a 12.5-mm orthoscopic ocular with cross hairs illuminated by a variable-brightness battery pack. It permits the photographer to select a star as a reference point for guiding. The guide star is kept cen-

tered on the cross hairs during the exposure. The standard Celestron reticle assembly is of 1¼-inch barrel diameter. For the Celestron 5, a removable adapter bushing is included to convert the ocular to 24.5-mm barrel diameter.

With the two accessories already discussed and one other, the lunar, solar, or planetary photographer can engage in the simplest form of deep sky photography—piggyback photography. In piggyback photography a camera with its normal taking lens is mounted atop the telescope. Typically 17-degree × 35-degree swaths of the night sky are recorded. A camera mounted in this fashion is shown in Fig. 12-11. The photographer guides through the main optics of the telescope in timed exposures ranging from 1 minute to a full hour. Celestron offers special piggyback camera mounts for each of their telescopes. The mount for the Celestron 14 is included in the purchase price of the base unit.

Piggyback photography is a good way to get started in deep sky photography for two reasons. First, there are celestial subjects that can be photographed only with a wide field, such as the star clouds of the Milky Way, long comet tails, and meteor showers. Second, piggyback photography with a wide-angle or normal lens is low-power

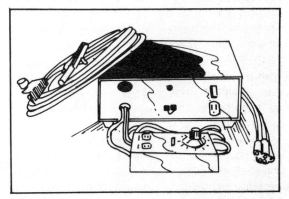

Fig. 12-10. The Celestron ac-dc drive corrector is utilized in deep sky photography.

Fig. 12-11. A camera is mounted on top of a telescope in piggyback photography.

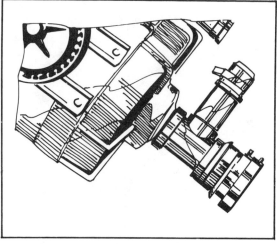

Fig. 12-12. The Celestron off-axis guider enables the coupling of a camera to the telescope.

photography and is relatively forgiving of guiding errors. The deep sky photographer who masters the short focal-length lens and the telephoto lens will be well-prepared for narrow-field, deep sky photography.

Having mastered wide-field or piggyback photography, you are now ready for narrow-field photography. It will be necessary to precisely guide the telescope during the photographic exposure. Celestron's off-axis guider is used to couple a 35-mm camera body to a Celestron telescope (Fig. 12-12). This will allow you to guide and photograph through the main optics of the telescope at the same time. The guider body accepts the illuminated-reticle ocular and employs a prism to divert light from a star that is off the edge of the photographic field into the guiding ocular. You can then select a suitable guide star.

While short focal-length piggyback photography is typically conducted at from less than 1 × (50-mm taking lens) with exposures ranging from about 5 to 30 minutes, Cassegrain-focus photography is conducted with the Celestron 5 at 25 ×, the Celestron 8 at 40 ×, and the Celestron 14 at 80

×. Exposures range from 5 minutes to an hour or possibly longer for very faint galaxies. To reduce the exposure times for the Celestron telescopes, the company offers some additional accessories.

Figure 12-13 shows the Celestron *Tele-Compressor*, which is a lens that reduces the effective focal length of the Celestron by one-half. The photographic speeds of the Celestron 5 and Celestron 8 are thus increased to f/5 and that of the

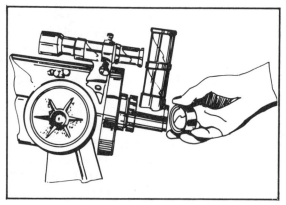

Fig. 12-13. The Celestron Tele-Compressor increases photographic speeds.

Celestron 14 to f/5.5. This fourfold increase in photographic speed means that exposure times can be cut to one-fourth. This effectively reduces the focal length and thereby the image scale. Also, the usable field at the film plane is reduced to about a 1-inch circle. The Tele-Compressor couples to the off-axis guider and accepts a T-ring and camera body.

The Celestron 5 guidescope is offered as an accessory for the Celestron 14 telescope. The guidescope includes a 5×24-mm finderscope, visual back, star diagonal, and illuminated-reticle ocular assembly. This add-on accessory also includes a tangent coupler assembly with manual slow-motions that permit you to sight in on stars up to 2½ degrees away from the optical axis of the telescope. Users of this guidescope are cautioned that the guiding-to-taking ratio becomes unfavorable, and photographs are much more susceptible to guiding errors than when using the off-axis guider and the main optics.

Astrophotographs of dim nebulous objects require a long exposure time, because most available films are relatively insensitive to faint light. These long exposure times necessitate correspondingly long (and tedious) periods of photographic guiding. Doubling the exposure time will not double the image density on the negative due to reciprocity failure in film. Fainter objects require a disproportionately longer exposure time to record satisfactorily on film. Color film is less sensitive to faint light than black-and-white film and requires a longer exposure time. The color balance of color film can change radically during long exposures.

The Celestron-Williams cold camera shown in Fig. 12-14 greatly increases film sensitivity (3 to 6 times for color film and up to 15 times for black-and-white film) and practically eliminates any shift in color balances. The cold camera increases film speed by chilling the film to subzero temperature during exposure, thereby greatly reducing reciprocity failure. It utilizes dry ice for cooling, which

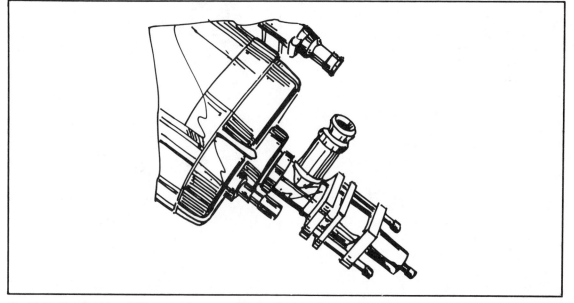

Fig. 12-14. The Celestron-Williams cold camera chills the film to subzero temperatures.

is available in 35-mm format only. The cold camera couples via the universal T-mount system to the Celestron off-axis guider.

Two isolator plugs, a focusing ocular assembly, a heater, operating instructions, and a carrying case are included with each cold camera. The cold camera uses a short length of film cut from a roll of ordinary 35-mm film (two recommended films are Ektachrome 200 for color work and Tri-X for black-and-white). Because the cold camera only accepts short pieces of film, be prepared to process the film yourself (commercial labs will not process short pieces of film). The cold camera has been used to obtain many beautiful photographs not possible otherwise.

PHOTOGRAPHY WITH THE
EDMUND SCIENTIFIC ASTROSCAN 2001

The Astroscan wide-field telescope will serve well for photographic purposes. Its relatively large light-gathering power provides telephoto capabilities under adverse lighting conditions that may preclude photography with slower lens-type telephoto systems.

When used in this manner, the lens of the camera is removed and replaced with a T-mount adapter specifically made for your particular camera. Adapters for popular SLR cameras are available through Edmund Scientific's catalog. A second adapter specifically designed for photography with the Edmund Astroscan 2001 is also required and is available as an optional accessory.

The photography adapter has male T-mount threads (42-mm×.75-mm) on one end. The other end has 1¼-inch outside diameter, or the same diameter as a standard eyepiece. Inside the photography adapter a negative achromat lens lengthens the optical path to bring the imaging plane back to the camera's film plane. This increases the effective focal length of the telescope and increased the f/ratio.

To use the telescope as a telephoto lens, re-

move the eyepiece. Slip the locking collar supplied with the adapter over the focusing tube. Remove the camera lens and replace it with the proper T-mount adapter for your particular camera. The photographic adapter is screwed into the camera adapter. The photographic adapter, with camera attached, is inserted into the focusing tube of the telescope.

After focusing the camera by turning the focusing knobs on the telescope, the focusing tube may be locked down in place by sliding the locking collar until it is resting on top of the focusing tube housing. Gently tighten the setscrew. A camera mounted in this manner is shown in Fig. 12-15.

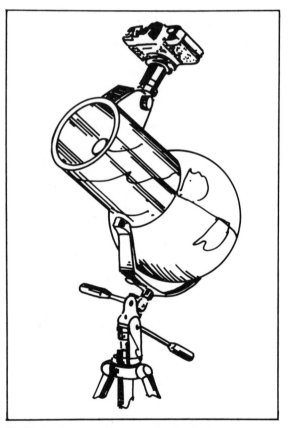

Fig. 12-15. A camera is mounted to the Astroscan 2001.

Fig. 12-16. The photographic Barlow lens adapter kit provides additional convenience in photography with the Astroscan 2001.

Edmund Scientific's photographic Barlow lens adapter kit will enable you to take photos through the Astroscan 2001 with the additional convenience of a detachable, high-quality, coated Barlow lens for double or triple power observing with your eyepiece. This kit is designed for use with a 35-mm SLR camera and will require an additional T-mount adapter for your particular camera (Fig. 12-16). This kit comes complete with a locking collar for camera focusing and detailed instructions. The kit is available with a precision Goodwin Barlow lens instead of the standard Barlow lens.

Another accessory that Edmund Scientific offers is their off-axis guider system. This single instrument eliminates the need for a guide telescope by employing a beam splitter that diverts a small portion of the light from the primary mirror. This diverted light is used to provide direct viewing

Fig. 2-17. Edmund's off-axis guider system (courtesy Edmund Scientific Company).

of the same image being recorded photographically, thus permitting exact tracking for long-exposure photography. By eliminating the guidescope, this system ends the problems of structural flexing between the main scope and guide. The system also includes an Edmund 28-mm RKE eyepiece, a 1¼-inch to 2-inch eyepiece adapter, a 2-inch outside diameter focus extension tube, and a camera-to-telescope adapter. This guider system is shown in Fig. 12-17.

DRIVE CORRECTORS FROM EDMUND SCIENTIFIC

Drive correctors provide the precision tracking necessary in some aspects of astrophotography. Edmund's right ascension drive is shown in Fig. 12-18. Fast and slow pushbuttons vary the tracking rate of the telescope in right ascension and produce precise tracking corrections during long exposures. This device will fit any telescope with synchronous drive motor up to 10 watts. It includes variable control for setting to solar, lunar, planetary, or sidereal tracking rates.

Figure 12-19 shows Edmund's dual axis drive corrector, which permits electronic corrections

Fig. 12-19. Edmund's dual axis drive corrector.

simultaneously on both telescope axes. The universal joystick remote controller allows the operator to use one hand for dual axis control, which makes for easy operation and convenience. This unit is designed to attach to Edmund's fork mounts and includes the declination drive motor, mounting brackets, and complete instructions on attachment and operation.

Edmund Scientific also offers the single axis drive corrector shown in Fig. 12-20. The single axis joystick control on this unit permits variable correction rates for variable right ascension control.

DEVELOPMENT OF THE CELESTRON SCHMIDT CAMERA

The tedium of early deep sky photography was due largely to the need for very long exposures. Exposures of 6 or 7 hours, which were practical because of the very dark skies, were not uncommon. Sometimes exposures of 10 or 11 hours, spread out over two or three days, were necessary. Such long exposures were necessary because

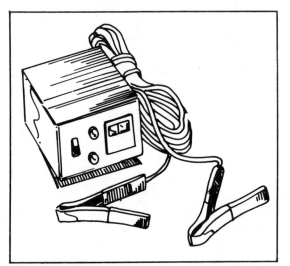

Fig. 12-18. Edmund's right ascension drive corrector.

Fig. 12-20. Edmund's single axis drive corrector.

photographic emulsions were slow, and most telescopes were of large focal ratio.

Even today a large focal ratio usually attends the long focal lengths needed for high-power studies of celestial objects, simply because large apertures are usually expensive. The result is to reduce photographic speed, requiring longer exposures to achieve a given image density on film.

The reduction in photographic speed may be acceptable when the subject is bright, as it is in lunar or planetary photography. When the subject is a faint nebula or galaxy, a slow photographic speed is usually unacceptable. The longer the exposure required, the greater the chance that atmospheric scintillation or guiding error will reduce the quality of the plate, if not render it completely unusable.

Another shortcoming in early deep sky photography was the need for numerous exposures to explore an extended nebula or region of the sky. This was partly due to the long focal lengths employed. A long focal length restricts field of view. There was a more important reason for this shortcoming, though, and it limited the progress of exploratory deep sky photography for two decades.

Most telescopes at the time were reflecting telescopes. They were easier and less expensive to make than refractors. Most of these reflectors were parabolic reflectors, because a parabolic mirror is the simplest mirror that can be used independently in a reflecting telescope.

A spherical mirror is simpler, but it is also unusable by itself on account of spherical aberration—the optical distortion in which light rays reflected from the outer zones of the mirror focus closer to the mirror than rays from the inner zones. This causes indistinct images over the entire field.

Deepening the curve of a spherical mirror into a paraboloid eliminates this aberration and produces a usable telescope mirror. The parabolic mirror produces sharp images only at the center of the field and at the expense of introducing even worse images at the edge of the field.

A point image near the edge of the field resembles a comet tail pointing outward with a parabolic mirror. (You can see why if you draw a parabola and trace the paths of parallel light rays that enter it at an angle to its axis.) This aberration is called *coma*, and it restricts the usable field severely within the limited field of a large focal-ratio mirror.

Why, then, couldn't the deep sky photographer simply have had his opticians make a mirror with a shorter focal length? Wouldn't this have increased the photographic field and the photographic speed? The answer is yes, but such a mirror is undesirable for two reasons.

First, a large-diameter mirror of short focal length is difficult to parabolize precisely. Second, and more importantly, the shorter the focal length, the worse the coma. Coma increases proportionally to field diameter and to the inverse square of the focal ratio.

Progress in exploratory deep sky photography

was thus hindered by the problem of coma. Not until 1931, when an obscure Estonian optician published a paper in *Zentral Zeitung fur Optik und Mechanik*, was a solution forthcoming.

The paper was entitled, "Ein Iichtstarkes komafreies Spiegelsystem." Its author, Bernhard Voldemar Schmidt, described how to construct a high-resolution, deep space camera of small focal ratio and wide photographic field that would produce images without coma.

Schmidt, who had joined the staff of Hamburg Observatory at Bergedort, Germany in 1926, built such a camera in 1930. Its aperture was 14 inches, its photographic speed was f/1.7, and its field was measured in degrees, not minutes of arc. The stellar images were pinpoint sharp from the one edge of the field to the other, and the nebulous images were bright and crisp.

Schmidt's invention was to revolutionize deep sky photography. Exposures requiring hours of tedious guiding, during which the astrophotographer was at the mercy of atmospheric scintillation, could be reduced to minutes. Entire fields of nebulae or galaxies could be photographed during a single exposure. The detail recorded in each object, on any part of the plate, would be of extraordinary quality and would depend less on film speed and more on the fineness of the film grain.

Bernhard Schmidt could not find a single buyer for his instrument. Not until the astronomers at what is now Hale Observatories saw Schmidt's photographs did it seem likely that a professional institution would fund the construction of a Schmidt camera.

In 1936, the year after Schmidt died, an 18-inch, f/2 Schmidt camera was operated at Mount Palomar. Astronomers discovered the first variable white dwarf stars, dwarf galaxies, and bridges of gas and dust-connecting galaxies with this instrument.

In 1949, the 48-inch f/2.5 Schmidt camera, which actually has an aperture of 49 inches, was operated at Palomar. This instrument mapped the celestial sphere north of declination −27 degrees—three-quarters of the heavens. The project took seven years.

In 1956 this sky survey was issued as the National Geographic Society-Palomar Observatory Sky Atlas. (The National Geographic Society had funded the project.) Eight hundred seventy-nine fields, each nearly 7 degrees square, were recorded, in both red and blue light, down to a limiting magnitude of 21.0 for stars and 19.5 for galaxies.

The 14-inch square contact prints are negative reproductions and must be examined with a hand magnifier. Any print in the Atlas may be purchased from the California Institute of Technology at a reasonable price. Compare the exposures made with a Celestron Schmidt camera and their corresponding exposures in the Atlas.

The Schmidt camera was a standard tool at many observatories by 1960. It was used to patrol for novae, track asteroids, to record the spectra of meteors or faint stars, and to photograph deep sky objects in various light wavelengths. In 1960 there were 15 Schmidt cameras with apertures greater than 23 inches, including the world's largest, the 53-inch camera at Tautenburg Observatory in East Germany. In 1971 Celestron introduced the camera as a moderately priced production.

Eliminating Spherical Aberration without Coma

A difficult challenge facing the pioneer deep sky photographer was to eliminate spherical aberration without the introduction of coma or some other aberration. Bernhard Schmidt met this challenge, and the logic of his achievement is illustrated in Fig. 12-21.

The top diagram illustrates the spherical aberration produced by a spherical mirror, in which three concentric light beams arrive at the mirror from infinity. (You are viewing the beams from the side.) Note that the rays defining these beams are reflected at different angles, and each beam has arrived at a different focus.

The greater the diameter of the outer beam, the greater the spread of focus between it and the inner beams. This spread of focus causes spherical aberration, and it occurs for light beams arriving at an angle and to those shown. Spherical aberration occurs uniformly over the entire field. How do we reduce or eliminate it?

You can place a mask with a circular opening in front of the mirror at its center of curvature. Placing this aperture stop ahead of the mirror reduces spherical aberration, because it restricts the diameter of the largest beam that can enter the system. The smaller the diameter of a beam, the less the spread of focus within it.

Because the stop is placed at the mirror's center of curvature, the spread of focus is reduced uniformly over the entire field. Every full-aperture beam that enters the system must strike the mirror at the same angle. Because all such beams are the same diameter, they must be reflected from the mirror to a common focal surface or a focal region of considerably diminished depth. See the middle diagram in Fig. 12-21.

The optical symmetry of this system means that any beam can be considered the on-axis beam. A line drawn through the center of the beam passes through the center of the stop, to the center of the portion of the mirror illuminated, then back on itself to the focal point. Spherical aberration is uniformly reduced over the entire field without introducing coma or any other off-axis aberration.

The focal surface is curved, so we can't use this system as a telescope. It can be used as a camera. Simply install a film holder that conforms the film to the focal surface. The focal surface is simply a section of a sphere, concentric with the mirror and bisecting its radius of curvature. This camera is the lensless Schmidt camera.

The lensless Schmidt can produce reasonably sharp images over the entire field. It can take advantage of the phenomenally wide fields of which the spherical mirror is capable. The lensless

Schmidt is not a true Schmidt camera. The camera is capable of only moderately fast photographic speeds (because of its aperature stop), and it does not eliminate spherical aberration.

To eliminate spherical aberration, the lensless Schmidt would require an aperture stop so small that the system would be f/10 or greater. Spherical aberration would then be undetectable, being less than Rayleigh's limit of resolution.

The main problem with a deep space camera would be its slow photographic speed. Because of the required focal ratio, a long focal length would be required for a large-aperture system. This would compromise field diameter.

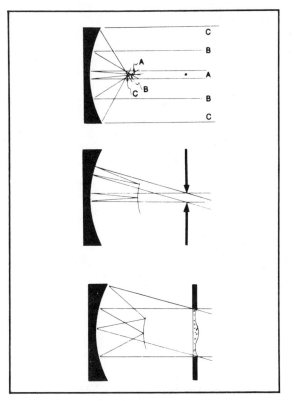

Fig. 12-21. This illustration depicts the logic behind Schmidt's camera (courtesy Celestron International, Torrance, Calitornia, U.S.A.).

For these and other reasons, Schmidt dismissed the lensless deep space camera as impractical. The spherical mirror was desirable, being so well-suited for short focal lengths and wide fields. The curved focal surface was acceptable, but the aperture stop had to go.

An unstopped system was needed that would bring all beams of light, whatever their diameter, to the same focal surface. Only this would simultaneously permit the large apertures, fast photographic speeds, and wide fields desired. Schmidt substituted a thin, zero-power, aspheric lens for the aperture stop of the lensless Schmidt. The figure of the lens is weakly convex at the center and weakly concave around the periphery, as shown in the bottom diagrams of Fig. 12-21. This lens can completely correct for the spherical aberration produced by a spherical mirror.

If a Schmidt corrector plate were placed ahead of the mirror in this diagram, it would, by refraction, diverge the outer rays, pass the intermediate rays undeflected, and converge the inner rays. All rays would be focused at the same point. As shown in the bottom diagram of Fig. 12-21, it would do the same for any beam of light of any diameter, on-axis or off.

With the invention of the Schmidt corrector, the images of numerous celestial objects could be recorded simultaneously with unprecedented resolution, brilliance and sharpness of field. The Schmidt-Cassegrain telescope would give razor-sharp images from one edge of the field to the other.

The Celestron Schmidt camera performs in the same manner as the giant Schmidt cameras at research facilities, but it is greatly simplified in operation and differs optically in one respect. The primary mirror of the Schmidt camera must theoretically be larger than the corrector plate by twice the width of the film holder to fully illuminate the photographic field. In practice a smaller primary is more than adequate for amateur and educational deep sky photography and many professional applications.

The primary mirror of the Celestron Schmidt camera is only slightly larger than the diameter of the corrector plate. The result is a vignetting or light drop-off at the edge of the photographic field, a consequence that is undetectable photographically and of importance only if you want to determine edge-of-field stellar magnitudes via densitometry.

Design

The design of the Celestron Schmidt is mechanically simple but beautiful. The mirror cell and film holder spider assembly are mounted to a zero-expansion Invar-bar cage that permanently holds these components to well within .001 inch of their correct positions. The factory-set focus of each instrument is permanently fixed, and the instrument never needs collimating.

The film holder, which is removable for film loading, is held on axis at the spider vanes by a magnet. Indexing pins machined to permit no lateral movement greater than .002 inch assure precise replacement of the holder in either vertical or horizontal format.

The resolution of the Celestron Schmidt camera at the film surface is film-grain limited. Resolution is about 1,300 lines per millimeter, as contrasted with the 40 lines per millimeter of an ASA 200 film or the 180 lines per millimeter of a microfile-type film. Considerable care should be taken in guiding to take advantage of the high optical resolution offered by the Celestron Schmidt. The size of a celestial object on film is often measured in hundredths of an inch.

The successful Schmidt camera photographer will select an electrically-driven equatorial mounting of superior quality and stability. A large aperture guide scope of diffraction-limited quality will be employed. Guiding tolerances will have to be observed carefully. At the image scale of the Celestron Schmidt camera, the smallest stellar image produced by most combinations of fine-grain film and developer is around 6 or 8 arc seconds. A guiding tolerance of 3 arc seconds is acceptable.

The unit of visual measure in the guidescope is the Airy disk. The apparent sizes of the disk in the Celestron 5, Celestron 8, and Celestron 14 are respectively about 1 second, ½ second, and ⅓ second of arc. A movement of about one, two, and four Airy disk diameters away from the cross hairs may be tolerated respectively.

A magnification of 100 × or so suffices to detect movements of this order. Using the Celestron illuminated reticle ocular assembly, you guide with the Celestron 5 at 100 ×, with the Celestron 8 at 160 ×, and with the Celestron 14 at 300 ×.

Image Defects on Film

Because of the Schmidt camera's optical and mechanical nature, it occasionally produces slight image defects on film. While these may be observed more clearly on the Palomar Observatory Sky Atlas prints, they will be noticeable on your negatives as:

☐ A pincushion-shaped image with some internal line structure found opposite the center of the negative from a bright star. This is a ghost image of the star produced by light reflecting from the emulsion, back through the system, and back to the emulsion again.

☐ A flare or streak originating at the edge of the field. This is caused by light reflecting from the inside wall of the film holder. The light originates from a stellar image lying just outside the field.

☐ Symmetrical streaks or "spikes" radiating outward from a bright star. The diffraction of starlight around the spider vanes of the camera causes these spikes.

Subjects for Photography

For wide-field celestial photography of high resolution, the Schmidt camera will provide quality photographs. You can capture on a single 35-mm negative any of the following with this camera:

☐ The entire North America nebula and its companion, the Pelican nebula, with their red fluorescence and ink-black patches of dust.

☐ Both halves of the Veil nebula, with their blue and rose-pink traceries set against the background of the Milky Way.

☐ All of the double cluster in Perseus, with its fiery orange supergiants scattered throughout and its hundreds of outliers.

☐ The complete spiral of Andromeda galaxy.

☐ The Orion nebula and the Horsehead nebula, with their dark clouds of obscuring matter etched sharply onto their pastel red, yellow, and blue luminosities.

☐ All of the Pleiades, blazing like sapphires and enveloped in blue, brushstroke nebulosity.

Not all deep sky objects are of such enormous angular extent as the Andromeda galaxy or the Rosette nebula. The Dumbbell nebula and the Whirlpool galaxy fit well within the typical telescope's narrow field. The fine details of such small objects are best photographed at a large, high-resolution image scale. Recording the fine details of deep sky objects like the Dumbbell or the Whirlpool through this type of telescope requires long exposures. Many fainter details go completely unrecorded due to reciprocity failure. The Schmidt camera, with its fast photographic speed and high resolution at the film surface, will record these faint details in minutes. The list of narrow-field subjects that can be recorded by the Schmidt camera is quite lengthy, but the company does not include such tiny objects as the Ring nebula or the Crab nebula. While the Celestron Schmidt will capture the prismatic colors of the Ring or the pastels of the Crab in minutes, the angular dimensions of these objects, plus their brightness, commend them to the telescope.

The astrophotographer should always seek:

—A focal length consistent with the size of the subject on film, the size of the grain, the seeing conditions, and the aperture of the telescope.

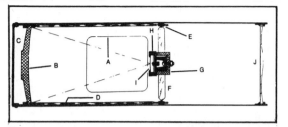

Fig. 12-22. Diagram of the Celestron Schmidt camera (courtesy Celestron International, Torrance, California, U.S.A.).

—A focal ratio consistent with the brightness of the subject, the speed of the film, and the range of available shutter speeds.

The size of a celestial object on film at prime, Newtonian, or Cassegrain focus is simply the tangent of the angular size of the object multiplied by the focal length.

Mounting

The Celestron Schmidt camera can be mounted on the Celestron telescopes or in their mounts (Fig. 12-22). They may, however, be mounted on any other telescope of similar size and stability. Fabricate adapter brackets to couple the preinstalled mounting brackets included with the camera to your telescope tube. The Schmidt camera can be purchased in the 5½-inch f/1.65, the 8-inch f/1.5, and the fork-mounted 8-inch configurations. The 5½-inch Schmidt camera is shown in Fig. 12-23 completely mounted. It is also available for mounting in piggyback style.

OPERATION OF THE SCHMIDT CAMERA

Schmidt photography is a form of deep sky photography that presumes a knowledge of celestial coordinates, polar alignment, setting circles, film, exposures, and guiding. Celestron includes this information in their telescope's operating manuals.

The field of the Celestron Schmidt camera makes it easy to acquire the target object. It may be desirable to accurately center a specific object. To quickly determine the exact part of the sky the camera is aimed at, load the film holder with plain white paper. When looking at the paper through the opened loading port, you can see the brightest stars.

Many people find it convenient to epoxy a small finderscope and bracket to the rear cell of the camera. When properly aligned with the Schmidt, the finder's entire field of view will be within the Schmidt camera's image area.

The film holder consists of a backplate with thumbscrew, a pressure plate, and an aperture plate (Fig. 12-24). The front of the pressure plate and the back of the aperture plate are precisely machined to match the radius of curvature of the camera's focal surface. The backplate of the film holder is also precisely machined specifically for each particular camera to assure that the film surface will lie exactly at that particular camera's focal surface. Film holders are not interchangeable between cameras, and extra holders for each should be ordered when the camera is ordered.

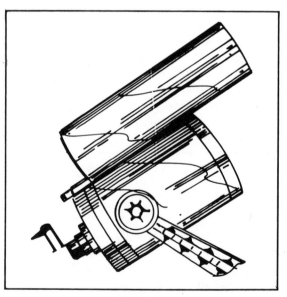

Fig. 12-23. The 5½-inch Schmidt camera is shown mounted.

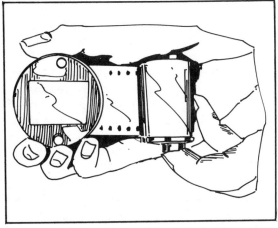

Fig. 12-24. The film holder consists of a backplate, pressure plate, and aperture plate.

Special film holders are also available from Celestron for filter photography. These holders, which are specific for the Kodak Wratten gelatin series, compensate for the displacement of focus due to refraction by the filter. They should also be ordered when the Schmidt camera is purchased.

To load the camera, remove the film holder from its magnetic saddle. Note the indexing holes that assure precise replacement of the holder even in the dark. Loosening the thumbscrew permits the film to be inserted between the aperture and pressure plates. Tightening the thumbscrew flexes the film to the proper radius of curvature. Tighten the screws manually. Too much pressure will push the film out ahead of focus. Check if the holder has a flat washer under the thumbscrew. This identifies a holder to be used for filter photography.

Care must be taken when loading and unloading the film holder not to damage the emulsion or expose it to light. All loading and unloading operations should be conducting delicately in a photographer's loading bag. The bag will contain the film holder with dust cover, a roll or cassette of film, a small pair of scissors, a film canister or developing tank, and a slip of soft paper. The paper should be the width of the film. It should be doubled over and be slightly longer than the aperture plate.

To load the film holder, remove the dust cover and loosen the thumbscrew. Unroll or pull out a length of film long enough to insert all the way into the folded piece of paper. Take care not to handle the emulsion side of the film. Slip the paper over the film. Snap off the film just beyond the paper. Insert the sandwich, emulsion side out, into the film holder and, holding a corner of the film, pull away the slip of paper. Tighten the thumbscrew, replace the dust cover, and the film holder is ready for the camera.

After the exposure, replace the dust cover. Place the holder in the loading bag and remove the cover. Loosen the thumbscrew, reinsert the paper, remove the film, and store it in the film canister or developing tank. If you store the film in a canister, the canister should be lined with soft paper.

Selecting a Film

It might appear that the best deep sky film would be the fastest film available. Because of reciprocity failure, though, high speed films may actually be slower than films with a lower ASA rating during a guided exposure.

In Schmidt photography, it is unnecessary to select a super-fast film because of the Schmidt camera fast photographic speed. A film of moderate speed, moderate-to-fine grain, and good reciprocity characteristics is required.

For close color photography with the Celestron Schmidt, the Ektachrome 64 or 200, or Fujichrome R-100, is suggested. The Kodak SO-115, H&W control, Kodak Plus-X, or the Kodak 103a series spectroscopic films are preferred for black-and-white work.

The 103a films are fast and grainy, but their low reciprocity failure makes them excellent for recording extremely faint nebulosities. Each 103a film is sensitive to a specific region or regions of the spectrum.

The 103a-E film is specific for red emissions such as those of the Lagoon nebula. The 103a-0 film is specific for blue emissions such as those of the Pleiades. The 103a-F is specific for both red and blue emissions and is a low-reciprocity, panchromatic film.

Plus-X is a good all-around deep sky film. It is well-suited for recording large or moderately large objects such as M31, M42, M8 or M20. H&W and SO-115 excel in capturing the details of resolvable clusters and bright to fairly bright nebulae and galaxies. H&W can be considered a fine-grain 103a-F in terms of sensitivity to the spectrum, and SO-115 can be considered a fine-grain 103a-E.

Making the Exposure

The exposure is begun and ended by removing and replacing a dark cardboard mask fashioned to hang from the front cell over the aperture entry. When the guide star is acquired and centered, the mask is removed and guiding commences. The mask is replaced when the exposure is finished.

The length of exposure required will be brief with the Celestron Schmidt—even for nebulous areas. The brightness of such areas at the focal surface depends on the square of the focal ratio. The image density recorded by an f/10 telescope in three-fourths of an hour will be matched by the f/1.5 Schmidt in 1 minute.

The precise length of exposure for any given camera will depend mainly on the sensitivity of the film and on the true apparent brightness of the celestial object. Because the brightness of resolvable stars at the focal plane depends only on the square of the aperture, star clusters are usually brighter than nebulae of the same magnitude. Magnitudes of nebulae and galaxies are listed as integrated magnitudes, so two such objects of the same magnitude may vary in apparent brightness. One may be larger than the other, as is the case with M51 and M81. The true apparent brightness of a nebula or galaxy may be indexed as its magnitude divided by its angular area.

Typical exposure times for the f/1.5 Schmidt are given in Table 12-1. To obtain the corresponding times for the f/1.65 camera, multiply by 1.2. The films are those recommended, and dark sky conditions are assumed. Starting time represents the mean exposure required to record the main features of all celestial showpieces. Optimum range represents the range of exposures recommended from very bright open clusters to faint nebulae and galaxies.

Exposures for the brighter nebulae and fainter star clusters will be somewhat less than those for faint nebulae and galaxies, and exposures for the brighter star clusters and globular clusters will be even less. The exposure times given in Table 12-1 are only guidelines. You will have to experiment to find the exposures that are just right for your particular site location, object, film, and camera. You might even want to exceed the optimum ranges by 5 minutes or more if the objective is to capture the faintest possible detail. If this is done, however, there will be decreased contrast between the subject and the night sky due to sky fog. This is the general background illumination of the night sky and is caused by the scattering of radiation from artificial lights, the scattering of sunlight in the atmosphere, and the glow of atmospheric molecules reacting to cosmic radiation. Your longest exposure will be limited by sky fog.

Table 12-1. Typical Exposure Times for the f/1.5 Schmidt Camera (courtesy Celestron International, Torrance, California, U.S.A.).

Film	Starting Point (mins.)	Optimum Range (mins.)
EKTA 200	8	2 - 10
EKTA 64	10	2 - 20
FUJI R-100	10	2 - 20
PLUS-X	8	2 - 10
H&W	10	2 - 15
SO-115	10	2 - 15
103a	6	2 - 6

Filter Photography

While intricately detailed exposures are achieved well before the sky fog limit, it is possible to push back the sky fog limit and achieve exposures of even more striking detail with filter photography. The radiation of wavelengths of sky fog lies mainly in the blue region of the spectrum, so it is possible with a red sensitive film and red filter to record red emission objects with greater contrast. Many of the brighter nebulae are red emission objects.

When a filter film holder is used, a longer exposure time is required than when using a standard film holder. With 103a-F film and a number 29 Kodak Wratten gelatin filter, the Celestron Schmidt has recorded the most delicate details of the North American nebula, the Pelican nebula, the Veil nebula, the Horsehead nebula, and Barnard's Loop in exposures ranging up to 30 minutes. This suggests that color exposures might also be extended slightly using a gelatin skylight filter. Filter photography requires a special film holder that Celestron has available as an optional accessory.

Hints for the Darkroom

Be prepared to process your own film when you use the Celestron Schmidt camera, but this should present no real difficulty. Even the processing of color transparencies requires little more than a developing tank, two or three chemicals, a floodlamp, and an hour's time. The following suggestions will aid you in the developing process:

☐ Your "plates" will be fairly small. In preparing them for agitation, make sure they are widely spearated on the developing tank's spool.

☐ If a color-reversal process is involved, be certain that the floodlamp illuminates each piece of film in its entirety.

☐ At the enlarger, in preparing composite prints for a mural, inspect the contrast between prints carefully. A second of printing time (and developing time) can make all the difference. The same will hold true for prints to be used in a sort of "sky atlas."

☐ If you envision great enlargement of your negatives, your standard 2-inch focal-length enlarging lens might be replaced with a used 8-mm camera lens of 15-mm focal length. This will make it easier to focus images of 50 × or larger on the baseboard.

☐ For studying negatives at enlargements of up to 100 ×, consider a quality microscope and a comparator reticle.

Caring for the Celestron Schmidt Camera

The components of the Celestron Schmidt camera are precisely machined, assembled, and rigidized for permanent focus and collimation. Certain precautions must be taken in transporting and handling this camera:

☐ When the instrument is not in use, remove the film holder and store it in the carrying case. When loading the camera, handle the film holder carefully to avoid dropping it on the mirror.

☐ The corrector may be removed for cleaning, but it and the surrounding shims must be replaced in exactly the same orientation and positions. The same surface of the corrector must face forward. Remove and index the shims before removing the corrector.

☐ When cleaning the primary mirror, do not pry the vanes of the spider assembly or attempt to remove or adjust any component of the Invar-bar cage.

Focusing the Celestron Schmidt Camera

You will never need to recollimate or refocus your Celestron Schmidt camera if you properly care for it. Factory refocusing and/or recollimation is available through Celestron for a reasonable price.

The optics of the Schmidt camera require:

☐ That the corrector plate be placed in front of the mirror at a distance equal to the radius of curvature of the mirror.

☐ That the film surface be placed exactly halfway

between mirror and corrector, with the surface of the film and the surface of the mirror concentric.
□ That the optical centers of the mirror and corrector and the mechanical center of the film holder lie on a line that defines the camera's optical axis.
□ That the mirror, corrector, and film surface be normal (untilted) with respect to this optical axis.

These requirements must be observed within limits of accuracy measured in fractions of .001 inch. The procedure for focusing and collimating a Schmidt camera must be conducted carefully.

Collimation means here that the optical center of each optical element lies on axis, and the element is normal with respect to the axis. Focus means the proper placement of these elements and the film holder along the optical axis, with the mirror and film surface normal to the axis. If a Schmidt camera is out of collimation, the Airy disks produced by hot spots on a target in the plate holder will appear flared or comatic, and uniformly so in the same direction over the entire field. If the camera is out of focus, the disks will simply be indistinct, but not necessarily uniformly so over the entire field. The center of the plate holder or film holder might lie exactly on axis at precisely its proper place, but the holder might be tilted with respect to the optical axis. The result would be sharp images at the center of the field and increasingly indistinct images out toward the edge of the field. Suppose the mirror is tilted. The images would be out of focus and comatic.

There are two ways to focus and collimate a Schmidt camera—visually and photographically. The size of the camera determines which method is the more practical, because the visual method requires a telescope of equal aperture and much longer focal length than the camera itself. The giant Schmidt cameras at the major observatories are thus focused and collimated photographically. The adjustment mechanism consists of a set of carefully calibrated micrometer screws coupled to the mirror. The astrophotographer makes a test exposure

of a given star field, reads the calibrations of the screws, changes the readings on the screws by given amounts, and makes a second exposure of the same star field. Comparing the two plates, he can perform the necessary computations to tell what further adjustments are necessary for the next test exposure, and so on, until the instrument is focused and collimated.

The Celestron Schmidt camera is focused in much the same manner, except the film holder is adjusted rather than the mirror. Because of the size of the Celestron Schmidt, you should use a telescope to very accurately observe collimation and focus at the focal surface. This method requires an aperture equal to that of the camera. The telescope's focal length should be at least five times longer than the camera's focal length. The procedure also requires two high-intensity light sources and a grating.

The telescope must be accurately focused at infinity prior to testing. Focusing on a star using a 12½-mm illuminated-reticle eyepiece to avoid accommodation focus of the eye will be sufficient. Because the testing procedure requires focusing away from and back to infinity, the position of the eyepiece or eyepiece assembly at infinity focus should be scribed on the eyepiece drawtube. The telescope and camera are placed front cell to front cell during testing. (The distance between the two is unimportant.) The testing then proceeds in three steps: the acquisition of rough focus, collimation, and fine focus.

To check for (or acquire) rough focus, a slip of white paper is placed in the film holder of the camera to serve as a target. A grating is placed at the focus of the telescope and illuminated with a high-intensity lamp. This produces an image of the grating on the target. By tilting the telescope or camera slightly, the image is brought to the target's center. The image of the grating is focused as sharply as possible on the target by adjusting the position of the film holder. This is accomplished by turning the pairs of spacer nuts holding the spider vanes on the three Invar bars. A quarter turn or less for each pair of nuts should be sufficient to achieve focus.

Do not loosen the nylon screws that center the Invar-bar cage inside the camera's tube assembly. Lateral movement of the cage defocuses the instrument.

Collimation and fine focus are checked after rough focus. This requires illuminating another type of target, which is placed in the film holder and illuminated by projecting a high-intensity light through the loading port of the camera. (The light should be shielded to prevent stray light from entering the telescope.) The target is a lithographic reproduction dot pattern printed on glossy stock. Illuminating the irregularities on the surface of the stock produces numerous hot spots that appear somewhat like stars through the telescope. The smallest of these form Airy disks. Collimation and fine focus can be checked by examining them.

To check collimation, select any hot spot on the target, bring it into sharpest focus, and examine it for coma or flare. If the image is comatic, the condition can be remedied by moving the corrector in the direction of the flare. Use shims at the edge of the corrector. If the coma persists even with this adjustment, the mirror may be out of tilt, possibly as a result of a sharp blow to the rear cell. The tube assembly of the Schmidt camera must then be re-seated against the rear-cell flange.

Checking fine focus is a somewhat more exacting process. If the Schmidt camera is in focus, then any Airy disk produced by a hot spot on any part of the target will be in sharp focus when observed through the telescope at infinity focus. Three reference focus points widely separated on the plate are sufficient to check focus. The three focus points should be located at the edge of the aperture plate where the three spider vanes terminate.

To locate the three focus points, the image of the grating used to acquire rough focus is projected onto the target. The position of the image indicates the position of the very small field that will be observed through the telescope. Focus at the center of the target should be checked first. To acquire focus, a hot spot below the resolution limit of the telescope is brought into the field. The telescope is focused until the Airy disk is in sharpest focus. The distance that the eyepiece must be moved from the infinity focus position indicates how far that part of the film holder must be moved.

Focal travel is proportional to the square of the focal length. If you are using the Celestron 8 telescope to focus the 8-inch Schmidt, this ratio is $(80/12)^2$. Move the eyepiece 0.25 inch in the drawtube to find sharpest focus at a given point on the target. That point must then be shifted 0.0056 inch. With sharp focus achieved for the center of the target, focus is then checked for each of the three reference focus points at the edge of the aperture plate where the spider vanes terminate. Patience is required. Moving the film holder at any of these points will tilt the holder with respect to the optical axis. Record your results for each progressive focus step.

To retain sharp focus at the center of the field while achieving focus at an edge-of-field point, the spider vanes opposite the focusing vane should be moved about half the distance the focusing vane was moved—in the opposite direction. The film holder may be reoriented in its saddle, and the three reference points can be examined again to further refine focus.

Chapter 13

Specialized Equipment

A TELESCOPE IS A VERY VERSATILE INSTRUMENT in itself, but with additional equipment attached to it, the possibilities are almost endless. This chapter will introduce you to many available accessories.

CELESTRON MULTIPURPOSE SCHMIDT TELEPHOTO LENSES

A *telephoto lens* is a camera lens that has a focal length longer than that of an ordinary camera lens. Although a telephoto lens will be quite useful in astrophotography, it will provide great detail and clarity when used with your telescope.

Most 35-mm cameras come equipped with lenses of focal lengths from 50 to 58 millimeters. These lenses produce the customarily small images with which we are all familiar. A bird at 50 feet will show up as no larger than a speck of dust, if at all. A telephoto lens will have a focal length ranging up to

1,000 mm or more. With such a lens, a bird at 50 feet will show up on film so large that it practically fills an entire 35-mm slide. This is the basic advantage of the telephoto lens. The photographer/viewer is capable of close-up observations and photographs of objects at a great distance.

The telephoto lens is capable of long distance candids and extremely selective focus because of its shallow depth of field. It can also eliminate distracting elements because of its narrow angular coverage. Another name for the telephoto lens is the long lens. The name derives from the fact that telephotos have long focal lengths. Traditional optical design means that with a long focal length comes a long lens housing. Telephotos often live up to their nickname literally and burden the user with excessive tube weight, tube length, and operational difficulties. Celestron offers two models: a 750-mm f/6 and a 1250-mm f/10. Both are very short on weight and bulk, and neither weighs quite 4 pounds nor exceeds 11 inches in length. They handle lenses that are only a fraction of their focal length easily.

The Celestron features the Schmidt-Cassegrain optical system that combines mirrors and lenses to fold a long optical path into a compact parcel and simultaneously produce the sharpest possible images over the widest flat field (Fig. 13-1). In this optical system the light (A) enters the system through a thin aspheric corrector lens (B). It is reflected by a large spherical primary mirror toward a convex secondary mirror, and (C) is reflected by the secondary mirror back through a hole in the primary mirror to the focal plane.

The Celestron 1250-mm f/10 telephoto is shown in Fig. 13-2. While the two telephotos offered by Celestron are the same type of lens and are both good for the same kind of photography or visual use, the differences in optical specifications between the two lenses tend to make each lens especially suitable for particular tasks. The two telephotos are virtually identical except for optics. The T-adapter for the 750-mm is shorter; thus, it is not interchangeable between the two.

The 750-mm f/6 is very "fast" for a mirror-lens telephoto making it especially suitable for focusing and shooting in dim light situations. The focal length of the optics makes it practical to hold this lens in your hand. Compared to the 1250-mm telephoto, the 750-mm has a closer near-focus distance

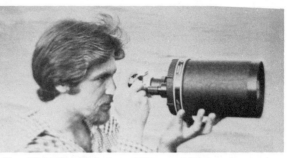

Fig. 13-2. The Celestron 1250-mm f/10 telephoto lens (courtesy Celestron International, Torrance, California, U.S.A.).

and a wider photographic and visual field. The low f/number and the wide field of the 750-mm make it particularly well-suited for certain types of visual observing whenever a very bright image and wide field of view are required.

The 1250-mm f/10 is a moderately fast lens of very high visual and photographic power. It is excellent for extreme close-ups, especially in typical daylight, whether the subject is at 20 feet or on the horizon. The 1250-mm has a slight edge over the 750-mm as a general-purpose daytime telescope. At night the 1250-mm is superior in revealing the details of the moon, planets, nebulae, and clusters of stars.

Mounting Your Camera

Virtually any 35-mm SLR camera with a removable lens and a focal-plane shutter can be coupled to one of the Celestron telephotos in seconds. Cameras with larger formats can also be coupled to this lens using custom adapters. To couple your 35-mm SLR to a Celestron telephoto lens, remove the lens of your camera. Attach the camera body to the T-adapter using the T-ring for your specific camera. The camera ring attaches to your camera just like your normal lenses and threads onto the appropriate T-adapter.

Thread the T-adapter over the rear-cell recess of the telephoto. You can position your camera body

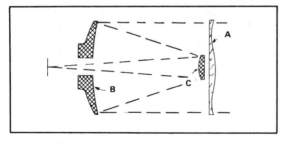

Fig. 13-1. The Schmidt-Cassegrain system combines both mirrors and lenses (courtesy Celestron International, Torrance, California, U.S.A.).

at the desired angle by loosening the knurled collar at the rear cell of the lens and rotating the entire camera body. Retighten the collar snugly.

While you can produce crisp photographs by holding a telephoto by hand, it is best to provide a sturdy support if you are using your lens for photography purposes. Telephoto lenses provide magnification with regard to the subject and also magnify any wavering of the lens during the exposure.

Focusing Your Telephoto

The focus control is located at the back casting. Turning the knob moves the primary mirror with respect to the secondary mirror and focuses the subject. You focus on close objects by turning the knob clockwise. Focus on faraway objects by turning the knob counterclockwise. Rapid shifts in focus may be required in some instances. Use a focusing magnifier for the most accurate focusing. Most camera manufacturers offer a 5 × or 6 × magnifier as an accessory.

Because the microprism or split-image range finder focusing aids in most cameras are designed for use with much faster (normal) lenses, they will appear to "black out" when used with long focal length telephotos. You must focus using the matte-screen area of your camera's focusing screen. If your camera has interchangeable focusing screens, check its instruction booklet to see if a special screen for telephoto lenses is offered. This screen may eliminate this effect. If you normally wear eyeglasses, be sure to wear them when focusing your lens.

If you intend to use the telephoto lens as a telescope, you might notice that the image shifts slightly when you are focusing at extremely high power. At about 200 power the shift might be as much as one-third the field. This is normal for the focusing mechanisms of these particular lenses.

Finding the Exposure

The telephotos offered by Celestron utilize the fixed-aperture system. You will be shooting at f/6, with the 750-mm and at f/10 with the 1250-mm. The only exceptions are the possible addition of a Tele-Extender or other similar device that will provide increased photographic magnification.

Because the aperture of your lens is fixed, the control of the exposure is effected by varying the shutter speed of the camera being used. If you own a light meter or if the camera has a built-in, through-the-lens light meter, this can be used to determine shutter speed. Some cameras have a special procedure for metering with nonautomatic lenses, so refer to the instruction manual provided with your particular camera for information. Fully automatic cameras may be used in the automatic mode if they have an aperture priority metering system (the camera selects the shutter speed).

If a separate light meter is used, remember that the lighting of the subject should be the same (or nearly the same) as the lighting being metered. Assuming the lighting is the same, set the meter for the aperture of the lens (a setting of f/11 for the 1250-mm or f/8 for the 750-mm will be close enough). Take a reading and adjust the shutter speed to what the meter indicates.

Because the acceptance angle of most light meters is much greater than the acceptance angle of either the 750-mm or 1250-mm lens, the meter may give an inaccurate reading. You will probably obtain more reliable readings by metering the incident light. If using this method, refer to the instruction manual provided with the light meter as to the proper method for measuring the incident light.

If a light meter isn't available, you can rely on the following rule of thumb. Using a film of a given ASA rating in typical daylight, the shutter speed at f/16 will be the inverse of the film's ASA ratings. If the film is rated ASA 64, then in typical daylight at f/16 the shutter speed will be 1/64 second or 1/60 second.

If you know what the shutter speed is at f/16, you can figure out what it is at f/11 or f/5.6 simply by knowing that opening up one aperture stop is equivalent to slowing down your shutter speed one increment. You can keep the exposure constant if you take one step up in shutter speed every time you open one stop. If 1/60 second is right for f/16, then 1/125 second is right for f/11, 1/250 second is right for f/8, and so on.

If you want to be sure of getting the shot, bracket your exposure with shutter speeds both faster and slower than that indicated by either your light meter or the rule of thumb. With practice, you can tell which shutter speeds (and films) are best for your particular subjects.

Photo Hints

The following hints on shooting with a long lens will be helpful if you are just starting out.

☐ Avoid shooting through mist, fog, haze, or heat waves. No telephoto lens can cut through these obstructions.

☐ Avoid shooting through window glass. Windows are not optical quality glass and can greatly degrade the image. Even shooting through an open window may ruin the sharpness of your photos if there is a noticeable temperature difference (heat waves).

☐ After any significant and rapid temperature change, allow 10 to 15 minutes for the lens to reach equilibrium before shooting. This is important during winter months when the outdoor temperature is low, and the lens has been stored indoors.

☐ Mount your lens on a sturdy tripod whenever possible. A tripod is probably the simplest and most effective way of improving the sharpness of your photos.

To obtain the sharpest possible photos (with the lens tripod-mounted), you might try using an air release cable (as opposed to the metal, spring-type cable) to trip the shutter. If it is practical, you can use the camera's self-timer to trip the shutter. Another way to minimize vibration is to manually retract the instant-return mirror of the SLR camera

being used prior to making the exposure.

Filter Photography

For telephotography under typical daylight conditions, you may sometimes use photographic filters to increase the contrast between subject and background and to gain a more natural rendition of colors. The Celestron Series VI drop-in filter set is designed to permit such flexibility in making exposures.

The filter set consists of six ring-mounted optical glass filters. These filters fit into the rear-cell recess at the back of the lens ahead of the T-adapter. They are held in place by the adapter when it is threaded onto the lens.

The number 1A is a "skylight" filter designed primarily for color work. It is salmon-pink in color and reduces the bluishness of shaded, overcast, distant, or aerial scenes. No exposure compensation is required.

The number 8 is a yellow filter for black-and-white work. It gives a good gray scale for natural clouds, sunsets, marine scenes, foliage, and portraits against the sky. Increase your exposure time by a factor of two.

The number 11 is a yellowish-green filter for black-and-white work. It lightens flowers and foliage, enhancing their texture in sunlight. The filter also darkens the sky background for portraits against the sky while generally yielding the most natural skin tone rendition with panchromatic black-and-white films. Increase exposure time by a factor of four.

The number 25, another filter for black-and-white work, is a red filter that produces spectacular cloud pictures, reduces haze in shots of distant landscapes, and dramatically emphasizes the texture of wood, stone, sand, and snow. Increase exposure time by a factor of eight.

The number 80A is a blue filter for color conversion. You can use it to expose daylight-type films to incandescent or photoflood lighting (3200 degrees Kelvin) indoors or out, avoiding the overall reddish cast that results without such filtration. For daylight photography, it produces interesting effects by emphasizing atmospheric haze and fog. Increase exposure by a factor of four.

The number 96 is a neutral-density filter for black-and-white or color work. Use it when you need to reduce exposures, such as when you are shooting a brilliant subject with a high-speed film, or when you want to decrease shutter speed to pan with a moving subject and produce a blurred background. Increase exposure time by a factor of two-and-a-half.

The listed filter factors apply if you are metering with a separate light meter. Behind-the-lens meters in most SLR cameras make it unnecessary to use the filter factors.

Filter photography is essential when shooting an aircraft or a flock of birds crossing the solar disk. Such shots are spectacular, but they are hazardous to your vision and your equipment without proper filtration.

You should never attempt to view or photograph the sun through any optical equipment without the proper solar filter, even when the sun is only moderately elevated above the horizon. The Celestron solar filter, will press-fit snugly over the front cell of the lens. The filter is an optically flat window coated with Inconel, a neutral-density substrate that reduces the intensity of the sun's rays to 1/100 of 1 percent of all wavelengths. This filter is available as a full-aperture, 5-inch model or as a stopped-down, 2-inch model. Both permit solar-disk photography or observation of objects such as sunspots, but the stopped-down model compromises your resolution somewhat and also reduces the effective f/value to f/15 or f/25.

Increasing Photographic Power

With Celestron's 750-mm telephoto lens attached to your camera, the photographic magnification is 15 ×, while that of the 1250-mm is 25 × (with 50 mm of focal length equal to 1 ×). There may be times when even greater focal lengths are desirable. An increase in magnification also results in an increase in detrimental effects of shutter vibration, atmospheric turbulence, haze, and wind on the sharpness of the pictures. As the focal length increases, the numerical f/ratio increases in the same proportion. Slower shutter speeds or faster, grainier films are necessary. Extended focal lengths are recommended only when no other alternative, such as closer approach to the subject, is possible.

A tele-converter is available at almost any photographic retail store. It converts a normal camera lens into a telephoto (or a telephoto into a more powerful optic). The device is typically 2 inches long or so and fits between the camera lens and body. The most common powers available are 2 × and 3 ×, although several suppliers offer intermediate ratios and even a few "zoom" models.

A tele-converter is an easy and inexpensive way of achieving modest increases in focal length with a telephoto lens. The Celestron 1200-mm telephoto becomes effectively a 2500-mm optic with a 2 ×. These devices generally work best with lenses of relatively high f/value. If you already own one, it will perform satisfactorily with these telephotos if it works well with your much faster ordinary camera lenses.

It may be desirable to increase the magnification by extreme factors. The photographic monitoring of destructive testing and the monitoring of hazardous manufacturing processes sometimes requires ultraclose inspection, as do some specialized forms of nature photography. Magnifications of up to several hundred times may be required. To meet these specialized needs, Celestron offers the Tele-Extender, which is a tube that lets you space back your camera body so you can project a highly magnified image to the film plane. The Tele-

Extender permits photographic magnifications of up to 650 × with the 750-mm lens and up to 1100 × with the 1250-mm lens.

The Tele-Extender tube attaches to the lens by the visual back accessory. The visual back threads onto the lens in the same way as the T-adapter. An ocular is placed into the visual back, and the Tele-Extender tube threads over the ocular onto the visual back. Your camera body is attached to the other end of the Tele-Extender with your camera T-ring.

The Tele-Extender may also be used without an ocular on the 750-mm or 1250-mm. Only a slight increase in focal length results. The 750-mm becomes a 950-mm f/11, and the 1250-mm becomes about a 1450-mm f/16. There is no reduction in sharpness.

The Telephoto Lens as a Telescope

The Celestron telephoto lenses are designed to be used visually and photographically. The two models are photographic versions of the Celestron Schmidt-Cassegrain telescope. With maximum magnification and steady air, you can read newsprint at a distance of nearly a mile, neon signs 10 miles away, or identify an airliner 20 miles away. The same near-focus capability that produces pictorial effects also works visually. Exploring your own backyard becomes an astonishing adventure in long-distance microscopy at or near the close-focus limit of either lens.

To use your lens as a telescope, you need Celestron's visual back, an ocular (eyepiece), and either a star diagonal or a porro prism. The visual back threads onto the rear cell of the lens. The star diagonal or porro prism is inserted into the back, and the ocular is inserted into the star diagonal or porro prism.

The star diagonal is a right-angle prism that reflects the optical path up to a convenient viewing position. This is the optimum configuration for stargazing with subjects overhead. Images are re-versed left-to-right, as in a mirror, with this accessory.

The porro prism is required for both erect and true left-to-right images. This is a straight-through viewing system that gives naturally oriented images for terrestrial viewing. The porro prism reduces image brightness slightly and is not recommended for stargazing, though the loss isn't noticeable in daylight. Figure 13-3 shows the Celestron telephoto equipped with star diagonal. The order is visual back (A), star diagonal (B), and ocular (C). The porro prism is used in Fig. 13-4, and the order is visual back (A), porro prism (B), and ocular (C).

The visual power of the Celestron telephoto depends on the focal length of the ocular selected for use with the lens. An ocular is generally named for its focal length. A 25-mm ocular has a focal length of 25 mm. To find the visual power of your lens, divide the focal length of the lens by the focal length of the ocular. Celestron offers a 2 × Barlow lens to extend the range of powers available with a given set of oculars. This accessory doubles the power of any of their oculars and also increases their eye-relief viewing distance. The Celestron

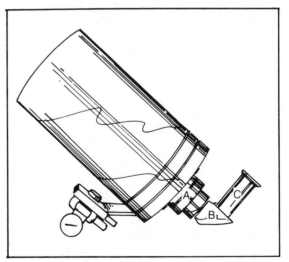

Fig. 13-3. The Celestron telephoto is equipped with their star diagonal.

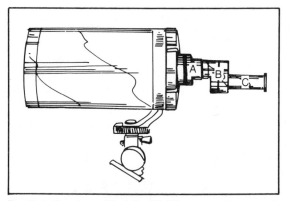

Fig. 13-4. Mounting will look like this if the porro prism is used.

Barlow inserts into the visual back and accepts either star diagonal or ocular.

The range of powers for the 750-mm telephoto lens offered by Celestron is from 30 × to about 200 ×. The range of powers for the 1250-mm is from 50 × to about 300 ×. In selecting oculars for use with these lenses, remember that high powers decrease image brightness, diminish the field of view, and magnify air turbulence.

The 750-mm performs best as a daytime telescope between 40 × and 60 ×, and the 1250-mm performs best between 50 × and 100 ×. A sparrow at near focus more than fills the field of view with the lower powers. You can inspect the structure of the sparrow's eye or study the facial features of a friend at half a mile with the higher powers.

The lower limit of useful daytime power for the 750-mm is given as 40 ×, even though the instrument's overall lower limit of power is specified as 30 ×. This limit also applies at night if you are viewing bright objects. It can be even higher if you are viewing particularly brilliant objects such as the full moon or a streetlight. The reason for this limit has to do with the interaction of mirror-lens optics and the eye's light-constricted pupil. If the pupil of your eye is smaller than the exit pupil (or emergent beam) of a mirror lens system, the housing of the secondary mirror may appear as a black shadow or

dot in the center of the field. Because the exit pupil increases in size with decreasing magnification, this is what happens when you use the 750-mm at less than 40 ×, and your eye has adapted to bright sunlight or to the image of the full moon. It is especially noticeable if you wear eyeglasses. If you wear glasses and detect a black shadow in the center of the field, try viewing without your glasses. If the dot persists, switch to higher power and leave your lowest power for stargazing. This shadowing is strictly a visual phenomenon and doesn't show up photographically.

Stargazing with the Celestron Telephotos

Low to intermediate powers are usually best for observing diffuse celestial objects such as star clusters, nebulae, or galaxies. Higher powers will be useful with moons and planets. The features of interest are tiny and bright and bear magnification well. Examine the object first with low power. Check Saturn's rings at 40 × or 50 ×. If the air is steady enough for a crisp image, use higher power. The hourly progression of celestial objects from east to west will be magnified in the same proportion as your target, so at very high powers your subject will literally appear to whisk across the field. The upper limit of usable power is imposed by your ability to "track" an object rather than the quality of your lens. See Tables 13-1 and 13-2.

CELESTRON RICH FIELD ADAPTER

The Celestron rich field adapter increases the brightness of faint deep sky objects and also the actual field of view on any of their telescopes. This adapter utilizes a high-quality positive achromat lens to compress the light cone exiting from the rear cell of a Celestron telescope. The rich field adapter will alter the optical properties of the telescope, so the f/ratio is one-half the normal f/ratio. You will then be using a telescope with an f/5.5 to f/5 focal ratio (depending on the Celestron tele-

Table 13-1. Information for the Celestron 750-mm f/6 Telephoto (courtesy Celestron International, Torrance, California, U.S.A.).

Celestron 750mm, f/6 Telephoto Reference Table

(Note: Field coverage is calculated for 35mm format only.)

T-Adapter Film Coverage (15x)

Angular Coverage (at infinity): 1.80° x 2.71°
f no.: 6
f no.: 7.9

Linear Coverage at Selected Distances

Distance	Coverage
15'	4.7" x 7.1"
30'	10.4" x 15.6"
50'	17.9" x 26.9"
100'	3.1' x 4.6'
500'	15.7' x 23.5'
1000 yds.	94.5' x 141.9'

Tele-Extender Powers, Focal Lengths, f/values, and Angular Coverage

Ocular	Photographic Power	Focal Length	f/ Value	Angular Coverage
25mm	90x	4,585mm	f/35	.300° x .450°
18mm	135x	6,660mm	f/55	.207° x .310°
12mm	210x	10,365mm	f/80	.133° x .199°
9mm	280x	14,065mm	f/115	.098° x .147°
6mm	430x	21,475mm	f/170	.064° x .096°
5mm	520x	25,920mm	f/205	.053° x .080°
4mm	650x	32,590mm	f/260	.042° x .063°

Tele-Extender Linear Film Coverage For Selected Distances

Ocular	15'	30'	50'	100'	500'	1000 yds.
25mm	.94" x 1.41"	1.88" x 2.83"	3.14" x 4.71"	6.28" x 9.42"	2.62' x 3.93'	15.7' x 23.6'
18mm	.65" x .97"	1.30" x 1.95"	2.16" x 3.24"	4.32" x 6.49"	1.80' x 2.70'	10.8' x 16.2'
12mm	.42" x .63"	.83" x 1.25"	1.39" x 2.08"	2.78" x 4.17"	1.16' x 1.74'	7.0' x 10.4'
9mm	.31" x .46"	.61" x .92"	1.02" x 1.54"	2.05" x 3.07"	10.2" x 15.4"	5.1' x 7.7'
6mm	.20" x .30"	.40" x .60"	.67" x 1.01"	1.34" x 2.01"	6.7" x 10.1"	3.4' x 5.0'
5mm	.17" x .25"	.33" x .50"	.56" x .83"	1.11" x 1.67"	5.6" x 8.3"	2.8' x 4.2'
4mm	.13" x .20"	.27" x .40"	.44" x .66"	.88" x 1.33"	4.4" x 6.6"	2.2' x 3.3'

Visual Powers and Fields of View
Linear Fields for Selected Distances

Ocular	20'	50'	100'	500'	1000 yds.
25mm	5.1"	13.8"	28.4"	12.1'	73.0'
18mm	4.1"	11.2"	23.0"	9.8'	59.0'
12mm	2.3"	6.3"	12.9"	5.5'	33.0'
9mm	2.2"	5.9"	12.1"	5.1'	31.0'
6mm	1.2"	3.2"	6.6"	2.8'	17.0'
5mm	1.0"	2.7"	5.5"	2.3'	14.0'
4mm	.8"	2.1"	4.3"	1.8'	11.0'

Visual Powers and Angular Fields

Ocular	Power	Field
25mm	30x	1.39°
18mm	40x	1.13°
12mm	60x	.63°
9mm	80x	.59°
6mm	125x	.32°
5mm	150x	.27°
4mm	190x	.21°

Table 13-2. Information for the Celestron 1250-mm f/10 Telephoto (courtesy Celestron International, Torrance, California, U.S.A.).

Celestron 1250mm, f/10 Telephoto   Reference Table

(Note: Field coverage is calculated for 35mm format only.)

T-Adapter Film Coverage (25x)

Angular Coverage (at infinity): 1.09° x 1.63°
f no.: 10
t no.: 12.5

Linear Coverage at Selected Distances

Distance	Coverage
20'	3.6" x 5.4"
30'	5.9" x 8.8"
50'	10.4" x 15.6"
100'	1.8' x 2.7'
500'	9.4' x 14.1'
1000 yds.	56.6' x 84.9'

Tele-Extender Powers, Focal Lengths, f/values, and Angular Coverage

Ocular	Photographic Power	Focal Length	f/ Value	Angular Coverage
25mm	155x	7,760mm	f/60	.177° x .266°
18mm	225x	11,275mm	f/90	.122° x .183°
12mm	350x	17,545mm	f/140	.078° x .118°
9mm	475x	23,820mm	f/190	.058° x .087°
6mm	725x	36,365mm	f/285	.038° x .057°
5mm	880x	43,890mm	f/345	.031° x .047°
4mm	1100x	55,180mm	f/435	.025° x .037°

Tele-Extender Linear Film Coverage For Selected Distances

Ocular	20'	30'	50'	100'	500'	1000 yds.
25mm	.74" x 1.11"	1.11" x 1.67"	1.86" x 2.78"	3.71" x 5.56"	1.55' x 2.32'	9.3' x 13.9'
18mm	.51" x .77"	.77" x 1.15"	1.28" x 1.92"	2.56" x 3.83"	1.06' x 1.60'	6.4' x 9.6'
12mm	.33" x .49"	.49" x .74"	.82" x 1.23"	1.64" x 2.46"	8.2" x 12.3"	4.1' x 6.2'
9mm	.24" x .36"	.36" x .54"	.60" x .91"	1.21" x 1.81"	6.0" x 9.1"	3.0' x 4.5'
6mm	.16" x .24"	.24" x .36"	.40" x .59"	.79" x 1.19"	4.0" x 5.9"	2.0' x 3.0'
5mm	.13" x .20"	.20" x .30"	.33" x .49"	.66" x .98"	3.3" x 4.9"	1.6' x 2.5'
4mm	.10" x .16"	.16" x .23"	.26" x .39"	.52" x .78"	2.6" x 3.9"	1.3' x 2.0'

Visual Powers and Fields of View

Visual Powers and Angular Fields

Ocular	Power	Field
25mm	50x	.84°
18mm	70x	.68°
12mm	105x	.38°
9mm	140x	.36°
6mm	210x	.20°
5mm	250x	.16°
4mm	315x	.13°

Linear Fields for Selected Distances

Ocular	20'	50'	100'	500'	1000 yds.
25mm	2.8"	8.0"	16.8"	7.3'	43.6'
18mm	2.3"	6.5"	13.6"	5.9'	35.2'
12mm	1.3"	3.6"	7.6"	3.3'	19.7'
9mm	1.2"	3.4"	7.1"	3.1'	18.5'
6mm	.7"	1.9"	3.9"	1.7'	10.2'
5mm	.5"	1.5"	3.2"	1.4'	8.4'
4mm	.4"	1.2"	2.5"	1.1'	6.6'

scope being used). This doubles the field of view for a given ocular and increases the brightness by a factor of four. There is a lower limit of useful magnification; this is where the exit pupil diameter exceeds the entry pupil of your dark adapted eye.

The Celestron rich field adapter is a visual model of the photographic Tele-Compressor. It allows right-angle viewing via a specially developed star diagonal, which is convenient when observing any object overhead. The rich field adapter consists of several components whose specifications and dimensions are critical. A special custom T-adapter (A) in Fig. 13-5 is used to mount the device on the back of a Celestron telescope. A Tele-Compressor lens (B) is threaded on the T-adapter. A custom-threaded ring (C) is attached to the Tele-Compressor lens. The 20-mm Erfle ocular (E) is inserted in the diagonal (D). The rich field adapter diagonal may be rotated to any convenient viewing position merely by loosening the slip ring, moving it to the desired position, and retightening the ring. Standard T-adapters and diagonals will not work very well, as too much of the field will be cut off. The new wide-field 20-mm Erfle ocular supplied with the rich field adapter will work well. The adapter is also fully compatible with the Celestron light pollution rejection (LPR) filter. The combination of the LPR filter and the rich field adapter will make for improved deep sky observing under less than ideal conditions.

You may use any 1¼-inch outside diameter ocular in the rich field adapter. The 20-mm Erfle ocular is also available individually. The wide apparent field of this ocular is approximately 60 degrees. It will work well on all Celestron telescopes and any other type of telescope with f/11 or lower ratios. The ocular is threaded to accepted filters and will also work with Celestron's Tele-Extender unit. This ocular is 1¼-inch outside diameter, but it can be used with .96-inch diagonals by using a special reducer bushing.

CELESTRON LIGHT POLLUTION REJECTION FILTER

The LPR filter blocks specific wavelengths of light primarily from mercury and sodium vapor lights (Fig. 13-6). It is not effective against moonlight. For best results with the LPR filter, use a

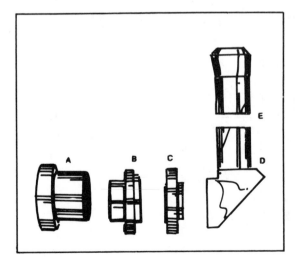

Fig. 13-5. A special custom T-adapter makes for simplified mounting.

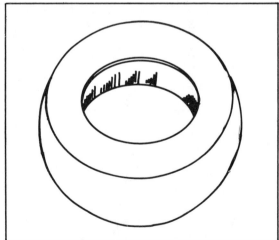

Fig. 13-6. The Celestron light pollution rejection filter blocks specific wavelengths of light.

low-power, wide-field ocular (as you would for deep sky observing without the filter). Because it is not effective against moonlight, observe with the filter when the moon is down. The LPR filter is also not effective in increasing the transparency of smog-polluted or cloudy skies.

The LPR filter has a slight amount of wedge inherent in the design. This creates a faint defocused ghost image on bright stars, but this is of no concern. The objects that this filter was designed to observe or photograph are comprised of faint stars. The eye is most efficient in detecting faint objects when it is fully dark-adapted. Because the eye won't become fully dark-adapted where there is a lot of light pollution, it will help if you shield your eye from extraneous light when observing.

When using the LPR filter for astrophotography, you will have to increase your exposure time by a factor of two or three to record stellar objects to the same density as without the LPR filter. Emission nebulae will be recorded to the same density on your film in approximately the same time with or without the LPR filter.

Due to the spectral response of a typical Celestron LPR filter, the light from an emission nebulae that is transmitted through to your film (or eye when used in a visual mode) is largely unchanged in its quality (spectrum). Color photography will yield an image of nearly natural color balance. Your choice of film may alter the color balance more than the LPR filter. Other apparent color changes can be caused by the lack of artificial radiation, background continuum, and minor nebular emissions that are filtered out by the LPR filter.

Some features of the LPR filter are:

☐ Makes bright light-polluted skies appear darker by rejecting radiation from mercury and sodium lights.
☐ Makes dark sky locations appear even blacker by rejecting unwanted natural light caused mainly by neutral oxygen emissions.

☐ Allows maximum transmission of the important wavelengths of hydrogen alpha, hydrogen beta, double ionized oxygen, and singly ionized nitrogen.
☐ Makes deep sky objects visible from city locations.
☐ Can significantly improve the views of galaxies and star clusters with associated nebulosity.
☐ Can be used in combination with the rich field adapter for deep sky viewing under less than ideal conditions.

CELESTRON DUAL-AXIS DRIVE CORRECTOR

A drive corrector is ideal for fine correcting in the position of your telescope for both photography and observing. The Celestron unit will electronically vary the speed (by changing the frequency of the supplied power) of the right ascension (east-west) motors, so you can overcome most causes of tracking error. A small motor allows corrections to be made in declination (north-south without touching the telescope. Both corrections are made with push button controls on a convenient hand-held control box.

Some features of the dual-axis drive corrector are:

☐ Operates on ac or dc. There is no need to purchase a separate dc inverter.
☐ Instantaneous corrections in right ascension, drive rate, and declination position.
☐ Provides variation in the overall average tracking rate. Differences in lunar, solar, planetary, and sidereal (stellar) tracking speeds can be made.
☐ Map light with rheostat dimmer control provided.
☐ On/off pilot light.
☐ Belt drive declination motor with mounting hardware.
☐ Dual speed declination motor.
☐ Declination motor operates independently of right ascension correction.

The drive corrector will provide a normal drive rate when the variable control on the handbox is set at zero. To slow down the drive, turn to a negative number; to speed up the drive, turn to a positive number. To set a sidereal, lunar or planetary drive rate, observe the object and adjust the variable control until the image drift stops. The fast or slow push buttons allow you to make the rapid drive corrections required during long exposure astrophotography. Each button is a momentary switch that must be depressed for as long as the correction is desired. When the button is released, the corrector returns to the drive rate selected by the variable control.

Many people find it helpful to adjust the variable control so the telescope tracking rate will be slightly slower than the rate of the object being photographed or observed. Only the fast button will need to be pressed (intermittently) to keep the guide star centered. This technique reduces the likelihood of accidentally depressing the wrong button during an exposure.

An outlet jack is provided on the front panel of the drive corrector for the 3-volt map light. This white light may also be used to light the reticle of the Celestron illuminated reticle ocular. The light intensity is easily adjusted by rotating the dimmer control.

CELESTRON DECLINATION ADJUSTMENT MOTOR

For Celestron owners with existing single axis drive correctors or telescope users who want the convenience of an electric motor on the declination (north-south) axis, Celestron offers the declination adjustment motor shown in Fig. 13-7. This motor with belt drive embodies a reversible, variable speed dc motor so that you can move your telescope in declination at the push of a button. The motor is self-contained and maintenance-free. It is controlled by a compact hand-control box that contains a 9-volt battery. This motorized declination adjust-

Fig. 13-7. The Celestron declination adjustment motor.

ment is the same motor as that supplied with the dual-axis drive corrector. This motor has two speeds: a fast slew (1 degree/minute) and a slower guiding rate (1/10 degree/minute).

Here are a few uses for the declination motor:

☐ During long-exposure deep sky photography, you may correct for residual declination drift if your polar axis alignment isn't perfect (poor alignment will make guiding more difficult and cause "field rotation" during long exposures).

☐ Piggyback astrophotography becomes a snap because you're photographing with short focal length lenses. You can "get away" with approximate polar alignment (point the polar axis at polaris) and correct for residual tracking errors with this declination motor and a drive corrector.

☐ The declination motor slew rate can be used to position your telescope for photographic work. It allows you to fine tune the object's position in your camera's viewfinder (or guide star in your off-axis guider) without annoying vibrations caused by shaky hands or bumping the telescope.

☐ Visual observations are greatly enhanced with a declination motor. As in astrophotography, the motor will allow vibration-free positional slewing. You are able to smoothly scan your telescope north and south across lunar crater fields at high powers without physically touching the telescope.

A declination adjustment motor is shown attached to the Celestron 8 telescope in Fig. 13-8. The unit comes with complete instructions for relatively simple installation. It can be purchased in coordinated colors to match those of your telescope.

CELESTRON TANGENT ASSEMBLY

The tangent assembly's basic function is to allow you to quickly and rigidly attach another optical instrument on top of a larger or primary Celestron telescope. It also allows you to point the two telescopes at the same or different objects. This is commonly referred to as independent targeting.

The most basic reason for attaching one telescope piggyback to another is to provide a smaller diameter guide telescope for the large instrument. This unit allows installation of a guide telescope (a telescope that you monitor the tracking accuracy with so you can apply corrective measures with the drive corrector), so you can photograph through one telescope and "guide" the exposure with the other. Due to the possibility of optical or mechanical flexure that no guide telescope can detect, an off-axis guiding system should be used for best results.

You can use a guide telescope to guide on the exact object you are photographing with a tangent assembly. You can also use the coupling system's positional adjustment to point the guide telescope in any direction (±5 degrees in right ascension and ±3 degrees in declination) to locate a bright guide star.

With the tangent assembly, it is possible for you to do astrophotography through a telescope that does not have a drive system by utilizing the drive and tracking systems of a larger telescope such as the Celestron 8, 11, or 14. You can use both instruments visually, photograph through the primary one only, or photograph through both simultaneously if you use the off-axis guiding system on one of the instruments.

Some features of the tangent assembly are:

☐ Easy installation on any Celestron telescope.
☐ No drilling or metal work required to achieve mounting.
☐ Allows independent pointing for both the primary and piggyback instrument.
☐ Allows attachment of any telescope/telephoto/camera with a tripod adapter and hole threaded for standard ½×20 attachment bolt.

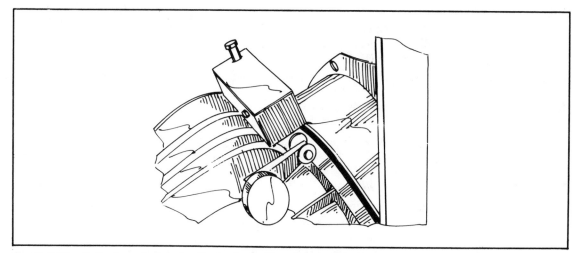

Fig. 13-8. The declination adjustment motor is attached to a Celestron 8 telescope.

CELESTRON OFF-AXIS GUIDING SYSTEM

An off-axis guiding system is designed for use in astrophotography. The Celestron system is composed of the guider body and the ocular assembly. A small precision prism in the off-axis guider body intercepts the image of a star at the edge of the telescope's field of view. Most of the image falls onto your film, but the single intercepted stellar image is projected onto the illuminated reticle ocular so that you can accurately guide your telescope during a long time-exposure photograph.

Features include':

☐ A red light emitting diode (LED) for reticle (cross hair) illumination.

☐ A sensitive rheostat to vary the LED's brightness.

☐ Use of an off-axis type guiding system totally eliminates the possibility of mechanical flexure that can ruin an astrophoto.

☐ A 12.5-mm orthoscopic ocular is used as the illuminated ocular.

☐ The cross hairs are inscribed on optical glass. They are not subject to damage by uniformed observers.

☐ The orange and brown battery pack is aluminum and will enhance the telescope's appearance.

☐ The supplied battery will last for years because of the low power required by the LED.

☐ The red LED that illuminates the reticle causes less eyestrain. It allows you to guide on fainter stars than would be possible with a white light.

☐ A 6-foot cord from the battery pack to the ocular allows you to set the battery pack where it is most convenient.

CLAVE PENTAPRISMATIC STAR DIAGONAL

The Clave line of telescopes and accessories from Cross Optics of Escandido, California are of high quality. Figure 3-9 shows the pentaprismatic star diagonal. For Cassegrain/refractor owners,

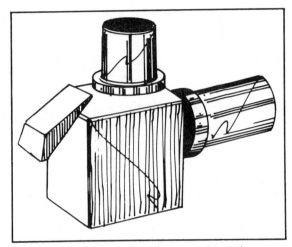
Fig. 13-9. The Clave pentaprismatic star diagonal.

this attachment directs the optical axis 90 degrees to one side by two total reflections within a pentaprism. Unlike other right angle devices that utilize one reflection and produce an "erect" but right-left reversed image, the Clave pentaprismatic diagonal maintains the astronomer's accustomed normal inverted image orientation. The image appears just as it would without this device, except that the eyepiece is in a convenient position for viewing with unstrained eye and neck muscles. Due to the optical tolerances maintained in the manufacture of this device, there is no loss of definition when using it. The observer will be able to view more comfortably for a longer period of time and with a familiar image orientation. The unit is designed for utilizing Clave Plössl eyepieces either in 27-mm or 1¼-inch barrel. The longest focal length eyepiece that can be used with unvignetted field is the 25-mm focal length eyepiece. This unit and the Herschel pentaprism have the 2 × Barlow lens built in. This lens may be removed for wide-field observation.

CLAVE HERSCHEL PENTAPRISM

This unit will allow observation of the sun

Fig. 13-10. The Clave Herschel pentaprism can be used to observe the sun directly.

directly in white light without filtration, while observing with the same convenience and image orientation as with the star diagonal discussed previously. Cassegrainian telescopes will require stopping down off-axis to avoid secondary mirror and support overheating. The unit is designed for utilizing Clave Plössl eyepieces in either 27-mm or 1¼-inch barrel. The longest focal length eyepiece that can be used with unvignetted field is the 25-mm focal length ocular.

CLAVE 2-INCH REFLECTING STAR DIAGONAL

This unit utilizes a quality optically plane mirror and one reflection to divert the optical axis 90 degrees. The mirror utilizes special enhanced coatings that result in 97-99 percent reflectance for maximum light efficiency. The mirror is extremely smooth without "micro-surface" irregularities. The overall figure indicates a convexity or concavity from edge to edge of less than 1/20 of sodium light. There is no detectable image degradation whatsoever. The Clave 2-inch diagonal has a working aperture of 48 mm, which is a bit more than most diagonals in this class. It will fully illuminate even the 65 and 75-mm focal length Clave Plössl oculars. This unit is designed for use only with the large 50-mm C.D. eyepieces (30-75-mm Plössls). It must be used with focusing mechanisms accepting 2-inch accessories.

CLAVE DEEP SKY FILTERS

These filters are designed to simply snap onto the eye end of any Clave eyepiece. Their deep sky filters are available as photovisual and visual. The photovisual filter optimizes transmission in the red spectrum where most nebula light is located, as well as the blue-green area where hydrogen-beta and doubly ionized oxygen and hydrogen-alpha are as high as 85 percent. Blocking transmission is approximately .02 percent, between 5400A and 6200A, which provides exceptional contrast. The photovisual filter makes possible the most rapid photographic exposures of any light pollution blocking interference filter made. The visual filter is especially well-suited for medium to highly light-polluted areas. It produces a beautifully striking image of emission nebulae, which will appear as celestial rubies when viewed through this filter.

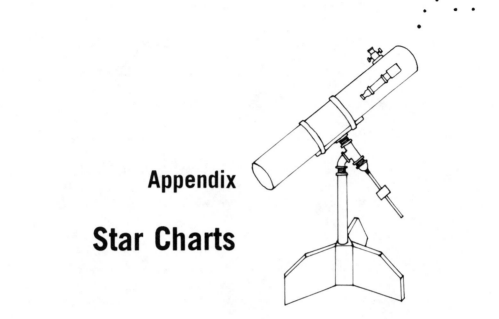

Appendix

Star Charts

THESE STAR CHARTS SHOW THE CONSTELLATIONS and identify the major stars. Figure A-1 shows the northern circumpolar stars that can be seen year-round from most areas of the United States. Figures A-2 through A-5 show how the heavens appear in the early evening in each of the four seasons. Figure A-6 shows the southern circumpolar stars. Many of these stars are only visible from the southern hemisphere.

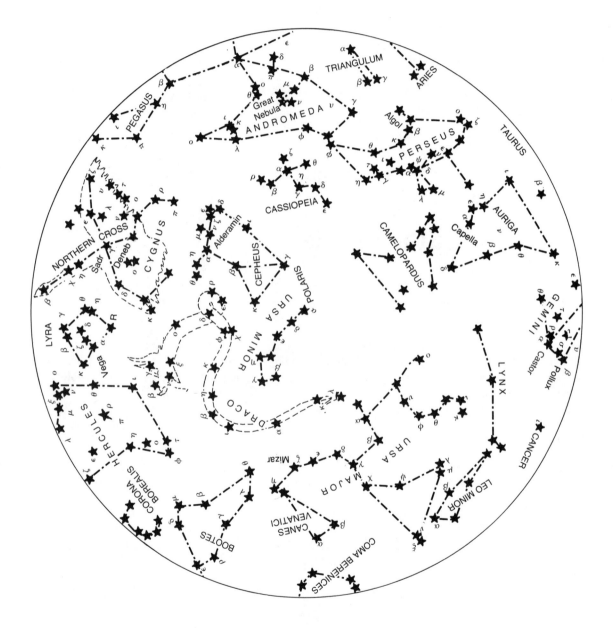

Fig. A-1. Northern circumpolar stars.

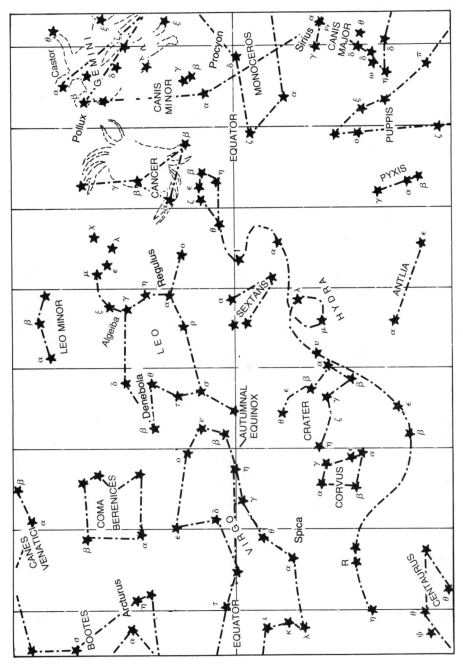

Fig. A-2. Stars of spring.

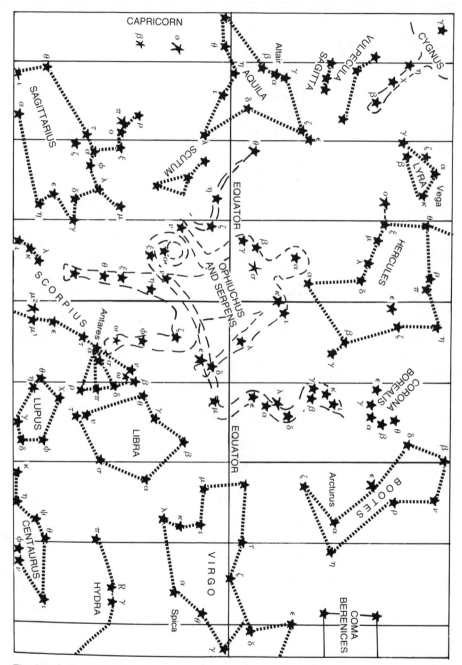

Fig. A-3. Stars of summer.

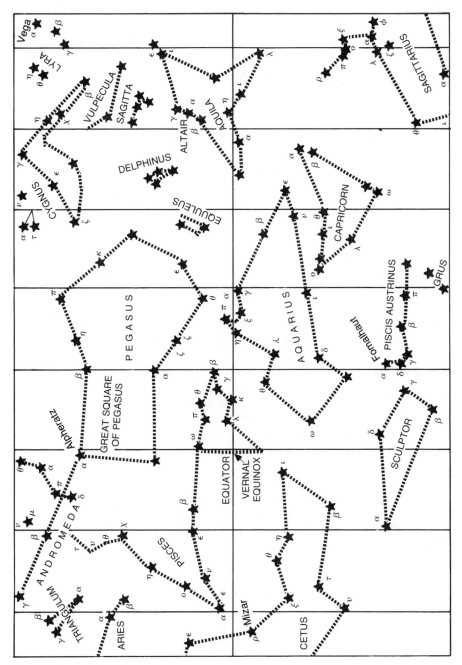

Fig. A-4. Stars of fall.

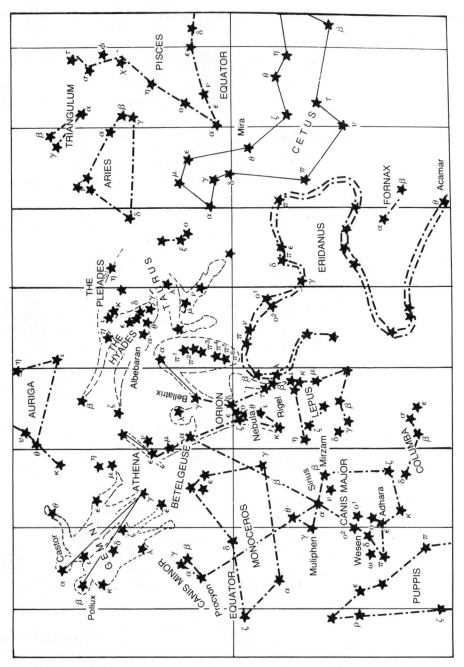

Fig. A-5. Stars of winter.

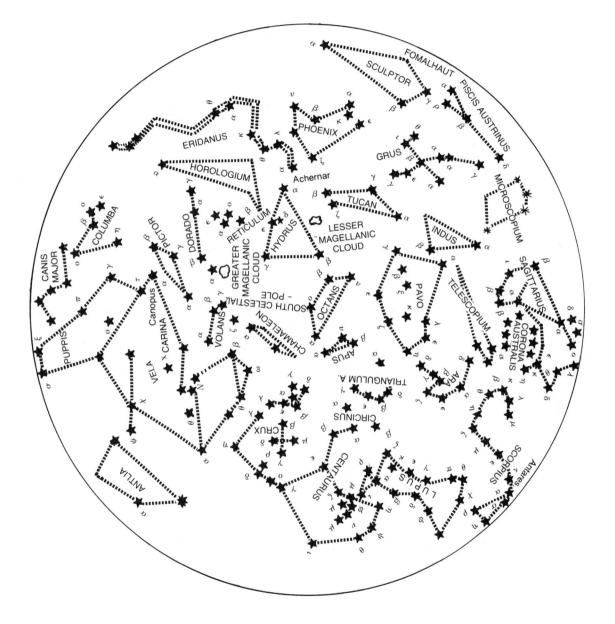

Fig. A-6. Southern circumpolar stars.

Glossary

achromatic lens—A lens arrangement commonly used in refractor telescopes whereby two different kinds of glass are pressed together to form a single lens. This has been found to significantly reduce chromatic aberration—one of the major problems associated with early refractors. The first primitive designs occurred in the late 1700s.

Airy disk—The image produced by an optical instrument of an object in space. When viewed through a telescope, the image will appear in the shape of a disk whose shape is formed by the reflected light rays coming to a single point.

aperture—The diameter of a mirror or objective lens in an optical instrument.

aphelion—The point in a celestial body's orbit when it is at its greatest distance from the sun.

asteroids—Sometimes referred to as minor planets. These celestial bodies maintain an orbit around the sun. The largest density of asteroids is found in the wide expanse of space between Mars and Jupiter. This area is known as the asteroid belt.

Barlow lens—A commonly used lens that consists of a concave lens inside a tube-shaped body. When placed in front of the objective lens, a Barlow lens will increase magnification significantly depending on the power of the lens.

binary stars—Two stars that revolve around a common center of gravity. A binary grouping may sometimes include three or more stars in relatively stable orbits revolving around the same center of gravity.

black dwarf—A late stage in the evolution of a star. When a star reaches this point, it has completely stopped radiating energy. No thermonuclear action is occurring at its core, because the hydrogen has been depleted. It takes a star billions of years to reach this point in its evolution.

Cassegrain telescope—A telescope whereby the optical system is composed of two curved mirrors—one short focal length parabolic mirror and one small hyperbolic secondary mirror. The telescope will be more compact, and the eyepiece will be in a more convenient position for ease of viewing. This is basically a modification to the Newtonian reflector telescope.

catadioptric telescope—A telescope that combines elements from reflector and refractor telescopes. It employs both a lens and a mirror, thus offering both long focal length and a compact, closed tube in a single instrument.

chromatic aberration—An effect normally associated with refractors whereby the white light is split up at different rates, thus causing false color images to appear on the outer edges of what is observed through the optics of the telescope.

collimation—The process of aligning the optics of a telescope.

coma—The head or nucleus of a comet that contains a high density of matter compressed tightly. This causes the comet to shine brightly. The coma may be extremely large. Some comets have diameters close to a million miles. It is also a distortion commonly associated with refractor telescopes that will cause stars near the outer edge of the field to appear blurred. This type of distortion will have the most serious effect on astrophotography.

comet—A body or mass composed of frozen particles that travels at extremely high speeds and in an eccentric orbit. As a comet reaches its nearest point to the sun, these frozen particles begin to warm up, causing the gases to spread out into a tail. As the comet leaves the close proximity of the sun, these gases fly off, causing meteor showers.

constellation—A grouping of stars in an identified and charted pattern. Many of the constellations that we know and observe today were originally named by early astronomers after animals, heroes, and even gods or goddesses.

definition—The accuracy of the images viewed through the opticals of a telescope over the whole field. The quality of definition will depend on both the quality of the mirror and its focal length.

diagonal mirror—In a reflecting telescope, the small, flat mirror that receives the rays from the primary mirror and transmits them at a right angle to the eyepiece. The diagonal mirror simply turns the rays to get them to the point where they can be observed through the eyepiece.

double star—Two stars in very nearly the same line of sight but seen as physically separate by means of a telescope.

eclipse—The effect caused when one celestial body is hidden from view by the body of another. It can also be caused by a body passing into the shadow of another.

equatorial mounting—A type of mounting for an optical instrument in which the main axis is parallel to the earth's axis. This method is much simpler for viewing stars, as they follow curved paths. It also has a second axis that is used for adjustments in the height of the celestial body being observed.

eyepiece—The lens or lenses that enlarge the image received either by the mirror or lens of the telescope so that the eye can see it.

focal length—The distance from the lens that the light rays will come to a point.

focal ratio—The focal ratio is determined by dividing the focal length by the diameter of the objective lens.

galactic clusters—Also commonly referred to as open star clusters. These are small, loosely grouped star clusters.

galactic nebulae—Gaseous, cloudlike, irregular areas in the Milky Way that are visible mainly due to the reflection from nearby stars. Some will appear very colorful and bright.

galaxy—An extremely large group of stars and other celestial bodies that are held together by gravitational force. Our own Milky Way is a galaxy.

globular cluster—Compared to galactic clusters, these groupings are extremely large and very dense.

gravitational force—That force that can either draw two bodies closer to one another or cause a central core of one body to draw more matter toward its center to form a tightly packed mass.

lens—A piece or pieces of glass sealed together to form a single lens that gathers light rays and bends them so they may be seen by the human eye.

light-year—A unit of length equal to the distance that light travels in one year in a vacuum or about 5,878,000,000,000 miles.

luminosity—The amount of radiance that a celestial body gives off.

magnification—In astronomy, the ability of a telescope or optical device to increase the size of the image a specified number of times for it to be viewed by the human eye.

magnitude, absolute—The brightness or luminosity of a celestial body if it was compared to other bodies being observed at the same distance.

magnitude, apparent—The brightness of luminosity of a celestial body as it appears to the observer from a point on earth.

Maksutov telescope—A telescope that combines properties of both reflecting and refracting instruments. It consists of a correcting plate that is spherical on both sides, with one side concave and one side convex. The convex part of the plate is the mirror side.

mass—In astronomy, the amount of matter found in a celestial body.

meteor—Also commonly referred to as shooting star. A meteor is really small particles of matter that heat up on entering the earth's atmosphere and appear as fiery streaks of light in our skies.

meteorite—A meteor that passes through the earth's atmosphere and strikes the earth's surface. They are usually composed of stone or iron.

meteor shower—A grouping of meteors that travel together and cross the earth's orbit at certain periods of the year at somewhat regular intervals. It is thought that these showers are the remains of former comets.

nebulae—A mass of gas and dust clouds that may be thin or dense. Some will appear as dark patches because there are no nearby stars to illuminate them. Others will appear quite bright and colorful.

Newtonian telescope—Named after Sir Isaac Newton. This reflector telescope utilizes mirrors.

nova—A star that suddenly increases its light output tremendously and then fades away to its former obscurity in a few months or years.

objective lens—The lens placed at the front of the tube in a refracting telescope. As light passes through this lens, it is bent to the focal point and then magnified by the eyepiece.

observatory—An observing site set up with optical instruments where astronomers can study the skies.

orbit—The path of a celestial body as it revolves around another one.

orthoscopic eyepiece—An eyepiece consisting of a triplet field lens and plano-convex eye lens. This eyepiece provides a flat field and clear image.

paraboloid—A figure having the shape of a parabola, which is a curve created by a plane slicing through a cone, with the plane parallel to the side of the cone. Mirrors used in reflecting telescopes are paraboloid.

perihelion—That point in a celestial body's orbit when it is at its closest distance to the sun.

primary mirror—The concave mirror in a reflecting telescope upon which the light rays form into an image. It is then reflected to the eyepiece for the observer to see once the image has been magnified.

prism—A wedge-shaped piece of glass that is used to bend the light rays. It may be used in a telescope instead of a flat diagonal mirror.

radiation—The radiant energy of a body that is transmitted through space as either a wave form or in particle form. This radiation makes it possible to learn about the properties of a celestial body.

reflector telescope—See *Newtonian telescope.*

refractor telescope—A type of telescope that utilizes an objective lens to receive light from a celestial body and bend this light to a point where it can be magnified and observed by the human eye.

resolution—In optical instruments, the ability to separate the objects in an image to the point where they can be seen individually. Resolution is invaluable in splitting double stars or binary stars.

reticle—A very thin piece of glass that is placed at a point in the focal plane of an eyepiece and provides assistance in centering the object being viewed in the field.

right ascension—The angle of a celestial body measured from the sun's position as it crosses over the celestial equator at vernal equinox. This system is the astronomical equivalent of longitude and can be measured in time and degrees.

rotation—The revolution of a celestial body on its axis. Earth rotates on its axis every 365 days.

satellite—A natural or man-made body that assumes an orbit around another body.

secondary mirror—The mirror that is located at a point near the eyepiece and reflects the image to the point where it can be magnified and made large enough for the eye to see.

shooting star—See *meteor.*

spectroscope—An instrument that utilizes a prism or prisms to split the white light into a rainbow of colors. By studying these rainbow colors, astronomers have been able to learn of what chemicals celestial bodies are made.

spectrum—The phenomenon whereby white light is split into many different colors.

spherical aberration—A distortion that can apply to either a spherical mirror or a paraboloid in which the rays deviate.

supernova—A very large star that has reached the stage in its evolution when it becomes so hot at its core that a huge explosion occurs.

variable star—A star whose brightness changes in fairly regular periods.

white dwarf—A star that is in one of the late evolutionary stages of its life. The star has cooled considerably due to the lack of hydrogen at its core. This causes the star to collapse. A star in this stage will emit little light, and its matter is very dense.

zodiac—An imaginary belt in the heavens that is divided into 12 constellations or astrological signs.

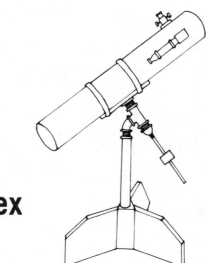

Index